Science Fiction Prototyping: Designing the Future with Science Fiction

Science Fiction Prototyping: Designing the Future with Science Fiction
Brian David Johnson

ISBN: 978-3-031-00668-5 paperback

ISBN: 978-3-031-01796-4 ebook

DOI: 10.1007/978-3-031-01796-4

A Publication in the Springer series

SYNTHESIS LECTURES ON COMPUTER SCIENCE #3

Lecture #3

Series ISSN Pending
ISSN 1932-1228   print
ISSN 1932-1686   electronic

Science Fiction Prototyping: Designing the Future with Science Fiction

Brian David Johnson
Intel Corporation

SYNTHESIS LECTURES ON COMPUTER SCIENCE #3

ABSTRACT

Science fiction is the playground of the imagination. If you are interested in science or fascinated with the future then science fiction is where you explore new ideas and let your dreams and nightmares duke it out on the safety of the page or screen. But what if we could use science fiction to do more than that? What if we could use science fiction based on science fact to not only imagine our future but develop new technologies and products? What if we could use stories, movies and comics as a kind of tool to explore the real world implications and uses of future technologies today?

Science Fiction Prototyping is a practical guide to using fiction as a way to imagine our future in a whole new way. Filled with history, real world examples and conversations with experts like best selling science fiction author Cory Doctorow, senior editor at Dark Horse Comics Chris Warner and Hollywood science expert Sidney Perkowitz, Science Fiction Prototyping will give you the tools you need to begin designing the future with science fiction.

The future is Brian David Johnson's business. As a futurist at Intel Corporation, his charter is to develop an actionable vision for computing in 2021. His work is called "future casting"—using ethnographic field studies, technology research, trend data, and even science fiction to create a pragmatic vision of consumers and computing. Johnson has been pioneering development in artificial intelligence, robotics, and reinventing TV. He speaks and writes extensively about future technologies in articles and scientific papers as well as science fiction short stories and novels (*Fake Plastic Love and Screen Future: The Future of Entertainment, Computing and the Devices We Love*). He has directed two feature films and is an illustrator and commissioned painter.

KEYWORDS

science fiction, futurism and future casting, user centered design, scenario planning, innovation, technology development and strategy, ethical implications of technology, research and development, human computer interaction, robotics and ai

Preface

Science fiction is the playground of your imagination. If you are interested in science or fascinated with the future then science fiction is where you can explore new ideas and let your dreams and nightmares duke it out on the safety of the page or screen. But what if we could use science fiction to do more than that? What if we could use science fiction based on science fact to not only imagine our future but develop new technologies and products? What if we could use stories, movies and comics as a kind of tool to explore the real world implications and uses of future technologies today?

This book is about science fiction prototypes (SF prototypes). SF prototypes are short stories, movies and comics that are created based on real science and technology. This is not a new idea, for over 100 years artists have been created fiction based on fact. What makes SF prototypes different is that they use these fictional creations explicitly as a step or input in the development process. Whether you're a designer, engineer, scientist, artist, student or strategic planner SF prototypes offer a way to imagine and envision the future on a whole new way.

I'm excited and honored that this book starts off with a Foreword from James Frenkel. If you wander down the aisles of any science fiction section or web site and scan the titles of the books you'll see a great many titles that have been touched by Frenkel but you'd never know it. James Frenkel is a senior editor at Tom Doherty Associates or as it's known to most SF geeks: Tor. Frenkel has shepherded and edited some of the greatest science fiction books and authors of the last thirty years. Now you may not be a SF fan but if you are you'll recognize names like Vernor Vinge, Joan D. Vinge, and Frederik Pohl, Andre Norton, Loren D. Estleman, Dan Simmons, Jack Williamson, Timothy Zahn, Marie Jakober and Greg Bear. And if you're not a SF fan then I'd recommend you start with any one of these legendary authors.

But beyond his talent as an editor, Frenkel's brain is a treasure trove of information, history and very strong opinions about all things science fiction. Every time I talk with him I'm amazed with his exhaustive knowledge of the people, stories and ideas that intermingle in the genre. He can draw from an early short story of Frederick Pohl in a discussion about human to robot interaction or use the best selling writing of Vernor Vinge to argue with you about the future of computing. He was the driving force and editor on the 2001 non-fiction collection *True Names: And the Opening of the Cyberspace Frontier*, proving that Frenkel is just as adept at exploring science fact as he is editing science fiction. In this book, we are going to explore short stories (Chapter 4), movies (Chapter 5) and comic books (Chapter 6) as different kinds of SF prototypes and how to build them. This book is much more of a "how to" book than a scholarly framework for design, research or engineering. Because the practice of SF prototyping is relatively new and is just beginning to be used in companies and taught at universities, we do not have the hard data and factual results needed to document the effect of SF prototyping on traditional design and engineering tasks. Never fear, this work is being done, but for the time being, we are more concerned about giving people a little background and grounding then getting them out there creating. Chapter 3 gives you a quick five-step process to beginning to build an outline for an SF prototype.

Throughout the book we are going to have a few conversations with some experts who have some interesting ideas about the intersection of science and fiction. We will talk about some history and look at examples of how people are envisioning the future. Science fiction writer and activist Cory Doctorow, talks to us about the history and power of science fiction and also how he puts together his highly popular short stories. Physicist, Emory University professor and writer, Sidney Perkowitz, discusses science and the movies. Sidney is an expert on Hollywood science, and we explore the science and message of *Moon*, the 2009 science fiction movie directed by Duncan Jones and starring Sam Rockwell. Finally, we talk about comics with a senior editor at the legendary Dark Horse Comics, Chris Warner. Chris has spent his life in the comic book industry, and we discover that comics might just be the best way to combine science fiction and science fact.

Throughout the book, we pull from a wide variety of other books, movies and articles, with the idea that if you see something you like or you find a bit of advice helpful, then you should seek out that book or author and dig in.

Finally, Chapter 7 looks at some concrete examples of how SF prototypes have been used in the development of AI and robotics. At the back of the book in the appendices, you will find the full SF prototype stories and scientific papers that we make reference to and use as examples throughout the book. Ultimately, as you read this book you will be able to build your own SF prototype, imagine your own possible futures and then go build them.

Foreword

Science fiction is a literature largely born in the 20th century. As Brian David Johnson details within these pages, it has its early roots earlier, in such works as Frankenstein by Mary Shelley and the works of Jules Verne and H.G. Wells. However, it was in the wildly developing technological cauldron of the 20th century that SF really got its true birth, in the pulps, grew and developed, matured and became the mature literature of a new millennium that it is today.

And now, Mr. Johnson proposes that science fiction can be a means by which scientists can explore the ramifications of new technologies, develop and test hypotheses, and find solutions to problems that come with pioneering techniques and emerging science.

Well, why not? Science fiction—whether in its text form, or film or comics—has, from the very beginning, been a literature of ideas. When the pulp magazines of the 1920s and 1930s were turning out a constant stream of two-fisted SF adventures, they were not quite the same as the twofisted detective stories, two-fisted war stories, two-fisted Westerns, two-fisted romances and other two-fisted tales of the pulps. Because SF was about the amazing, unimaginably exciting future.

Like comics, like science fiction movies before *2001: A Space Odyssey*, science fiction stories were considered lowbrow stuff, the kind of thing that kids like me—and probably every other kid who read SF and comics and loved the tacky but exciting SF flicks of the 1950s and 1960s—were not supposed to read or see. They were, as Brian points out later in the book, not healthy activities for young minds.

Aside from the paranoid and misguided warnings of Frederik Wertham's *Seduction of the Innocents*, I am not sure exactly what our parents thought would happen to us when we did these things. What I do know is that the combination of science and science fiction was positively intoxicating to our eager young minds. Within the ranks of science fiction readers and others interested in the field, there has long been a debate: what is the golden age of science fiction? Some argue for the 1930s, others the 1940s, or the 1950s, 1960s or 1970s. But most people will agree that in truth, the golden age of science fiction is around twelve—or whatever the age at which someone first becomes hooked on the stuff.

In the early 1960s, as the space program of the United States was getting started, one could see that the designs of spacecraft and other hardware bore a resemblance to the ideas and illustrations of science fiction books, comics and films. It was not hard to understand how that could happen. A lot of the scientists, engineers and technicians working on the Mercury, Gemini and Apollo programs had themselves grown up on that crazy skiffy stuff.

The lure of the future—of the challenge of being part of creating the future—was, and remains a potent lure for people of vision. The visions have changed over the decades, but the challenge and the allure of creating something new, something that nobody has created before, is something that drives those who create science, as it does those who create science fiction.

Within these pages, Brian David Johnson has brought together the talents of a number of very talented people to show, through interviews with professionals in the three fields of fictional creation— science fiction stories, films and comics—and also professionals in the sciences, how science fiction can be used to prototype methods to make science and technology and solve problems that occur in its development.

But also within these pages is much more. There are nuggets of wisdom and insight from such people as Cory Doctorow, Gregory Benford, Will Eisner, Peter David and other very talented, creative people, about the craft of writing, the structure of fiction and the history of the genre. There are also tales here—real-life tales—of some of the great legends of these fields, including one of the most iconic science-fiction writers of the 20th century, Isaac Asimov and the legendary, towering genius of comics, Will Eisner, creator of *The*

Spirit and a master of the form acknowledged by comics writers and artists alike as a seminal influence on the modern form.

The history in these pages isn't dry or dusty—it's exciting and dynamic, because it all deals with the ways in which brilliant, creative people have helped to change and develop the modern forms that are the focus and basis for the science prototyping that this is all about. You'll discover a good deal about the Silver Age of Comics of the late 1950s and early 1960s, when Julius Schwartz breathed new life into DC comics with reimaginings of characters like The Flash, Green Lantern and others, and Stan Lee seemed to have created the Modern Marvel Comics of Spiderman, The Fantastic Four, the Incredible Hulk and others.

Doubtless I'm biased, because I was a kid when that all happened, but I have the feeling that it's not mere coincidence that some of the characters who were brought to life then have become great film franchises, like the Uncanny X-Men, Spidey himself and Iron Man, for starters.

But this book is also about science, and anecdotes about Albert Einstein, Steven Hawking and others also enliven the text.

And yet, with all of this cool stuff, the heart of the book, is still Brian's how-to, step-by-step tutorials on how to use the three art forms of science fiction in the service of developing scientific solutions to problems that require creative thinking, with what pop culture today terms "outside-the-box" thinking.

There are some special bonuses here as well, including a full-color comic by the legendary Steve Ditko.

Science fiction writers famously deny that they predict the future, and I must agree with that. SF is, first and foremost, fiction, not science. When a science fiction scenario is mirrored by a scientific development in the real world, it is a happy coincidence, no matter how much fiction might have inspired that development. But I predict that if you read on, you will not only learn a great deal; you will also enjoy yourself in the process.

James Frenkel

Productively confusing science fact and science fiction may be the only way for the science of fact to reach beyond itself and achieve more than incremental forms of innovation.

Julian Bleecker
Artist and Technologist
Author of *Design Fiction*

Without science there is no science fiction.

Michio Kaku
Theoretical Physicist
Author of *Physics of the Impossible*

Dedication

For Vic, Simon, Sarah, and Sumi

Acknowledgments

The story of SF prototypes really began several years ago in Ulm Germany during a late night conversation with Vic Callaghan and Michael Gardner. The spirited debate continued as we walked home under the looming spire of the Ulm Munster—I cannot thank these two gentlemen enough for their good humor and intellectual curiosity. And Michael doesn't even like SF!

Part of the joy of writing this book was having some really compelling and fascinating conversations with Cory Doctorow, Sidney Perkowitz and Chris Warner. I consider my discussions with these three to be a privilege and I will always be indebted to them.

This book is filled with references and quotes from numerous authors, scientists and researchers. SF prototypes would not be possible without their imagination and intelligence: Brian Aldiss, David Wingrove, Michael Ashley, Isaac Asimov, Stanley Asimov, Kelley Baker, John Baxter, Gregory Benford, Elizabeth Malartre, Julian Bleecker, Nick Bostrom, John Brady, Michael Brooks, Arthur C. Clarke, Frank Close, Peter David, Karl Schroeder, Paul Dourish, Genevieve Bell, Simon Egerton, Graham Clarke, Victor Zamudio, Will Eisner, Russell Evans, Syd Field, Gardner Fox, Ron Goulart, Lois Gresh, Robert Weinberg, Stephen Hawking, Sarah Perez-Kris, T. D. Ferro, J. R. Porter, Kar-Seng Loke, Scott McCloud, Alan Moore, Steven Schneider, Nathan Shedroff, Chris Noessell, Mary Shelley, Alan Stelle, H.G. Wells and Paula Zizzi.

Over the years SF prototyping has seen the support of some incredible people and I don't think we would have made it this far without them: Justin Rattner, Tadayoshi Kohno, Sumi Helal, Duckki Lee, Wolfgang Minker, Michael Weber, Hani Hagras, Achilles D. Kameas, Juan Carlos Augusto, Jeannette Chin, Don Wallace, April Miller, Antonio Tatum, Jim Olsen, Klaus Obermaier, Sean Hanna, Darrin Johnson and Vernor Vinge.

I want acknowledge the University of Washington and Professor Sarah Perez-Kriz's "Science Fiction Prototyping" class for piloting this book and developing some thoughtful and engaging SF prototypes.

Thanks to Mike Morgan for his courage to publish this book and his enthusiasm for its rather unconventional subject matter.

Sandy Winkelman takes my words and turns them into not just pictures but whole worlds—my collaboration with him has been incredibly important to me and I could never thank him enough 25.

Contents

CHAPTER 1

The Future Is in Your Hands

In 1983, I was 11 years old. That year, I saw a movie that changed my life forever. That movie affected me deeply that there was one point during the show that I got so into the story, so wrapped up in the drama that I had to leave the theater and walk around in the lobby a little to calm down. Now for an 11-year-old … that was a movie!

My memory of that movie and the ideas it put in my head still affect me today. This might sound a little overblown but it is true. Now the title of the movie might not be what you would expect and how it affected me might seem even less obvious, but they are both significant.

The title of the movie was *WarGames*. Starring Mathew Broderick and Ally Sheedy, the movie tells the story of David Lightman (Broderick), an American teenager who is really good at computers but not so good at life. Essentially, he was a computer geek before most people really knew what a computer geek was. In the movie, David accidentally hacks into the government's War Operations Plan Response computer (WOPR) while he is looking for a mysterious game company called Protovision. Protovision's sleek black brochure pictures kids, their faces aglow with wonder and excitement with the text: "Things will never be the same … A quantum leap in computer games from Protovision." Who would not want to play those games? That was the mystery and wonder that *WarGames* was tapping into. Things would never be the same again, and it was computers that were going to bring it all to you.

As the movie moves forward, we learn that the artificial intelligence (AI) that runs WOPRs was designed by Stephen Falken, a character loosely based on the theoretical physicist, Stephen

Hawking. Falken has named the AI, Joshua, after his dead son, and teaches the computer through a series of strategy games like tic-tac-toe, checkers, poker and chess.

When David hacks into WOPR, he sees a series of tactical games like Theaterwide Biotoxic Chemical Warfare and Global Thermonuclear War. Thinking he has found the game company, he begins playing Global Thermonuclear War with Joshua. What David does not know is that WOPR has taken over the United States nuclear missile system and is preparing to launch a strike against the Soviet Union and start World War III.

In the early 1980s, the Soviet Union was still very much a threat, and the specter of nuclear war hung over both our countries. With the popularity of personal computers, like the Texas Instruments TI99, the concept of computers was taking hold in American pop culture. Ultimately, *WarGames* is a cautionary tale about the futility of war and the danger associated with giving computers too much control over our lives. But *WarGames* was just one of the movies from the 1980s that capitalized on the growing personal computer craze. Movies like *Tron, D.A.R.Y.L., Weird Science, Electric Dreams, The Last Starfighter*, and *Explorers* began to tell stories that brought computers into our homes and our daily lives. For the first time, science fiction was coming into your house and you could be the main character. But it was not science fiction anymore … it was real. The computers were real, the technology was real, and you could program your own computer to do almost anything it seemed. With imaginations fueled by these future visions, an entire generation started programming, building games and basically geeked out doing all the things that today, in 2011, seem as normal as flipping on the light switch. Today, most of us live quite comfortably with computers knit into everything we do, but back in 1983, it was new and exciting.

So at this point, you might be wondering, how did this movie change my life?

Well, during the movie when David is hacking into WOPR and playing games, the film cuts between the ominous blinking activity lights on WOPR and David playing war games with Joshua. Staring at those lights in the midst of the action, I wanted to know what was going on inside the computer. I tried to imagine how the computer worked. I wanted to know how an AI that complex could be built. How did you program such a thing? With my heart racing, I imagined how to do it. I could envision what was going on inside the big box with the silly name on the side.

I was already a geek before I saw *WarGames*, but the movie showed me that computers could be exciting. For me, the computer, Joshua, was the hero of the movie. I did not really care what happened to David. To me, *WarGames* was exciting because for the first time I imagined how to build something that I could not see. My mind literally opened up in that dark theater. I know now that I was thinking through high level system architectures, software stacks and network diagrams. The images of the WOPR and the AI were complex and intriguing, and if I used imagination and what I knew about computers, I could see how you might build it.

The most famous line from the movie is when the AI Joshua asks David, "Shall we play a game?" the words crudely vocalized through a make-believe voice synthesizer. But those words "Shall we play a game?" captured everything that was exciting and amazing about the movie.

Yes! I wanted to play a game. And I have been working with science and imaging technology ever since.

WarGames AS AN SF PROTOTYPE

In 1983, *WarGames* was a thriller of near-term science fiction. The writers, Lawrence Lasker and Walter F. Parkes, with the help of Peter Schwartz from the Stanford Research Institute, combined the capabilities of a new kind of personal computer with the threat of a nuclear war brought on by over-automation. When you think about it that way, *WarGames* is a kind of science fiction (SF) prototype.

What is a science fiction prototype? Stated simply, it is a short story, movie or comic based specifically on a science fact for the purpose of exploring the implications, effects and ramifications of that science or technology (we'll get into the details of SF prototypes and prototypes in general in Chapter 2). *WarGames* imagines what could happen if a computer was given control over the U.S. government's missile defense system, and something went terribly wrong.

Of course in the movie, the terribly wrong things are that David stumbles onto Joshua, the AI, and the two begin playing war games, not knowing that the games are being played out in real life.

I think what captivated me at 11 was that the movie allowed me to imagine the implications of a future that seemed very real. I knew enough about computers to know that this *could* possibly happen but what really captivated me was imagining *how* it could possibly happen.

The goal of SF prototypes is to start a conversation about technology and the future. This is incredibly important because the future is not set. The effects of science and technology are not a predetermined thing of nature. The future is made everyday by the action of people. We control our own future. It is precisely because of this that we must talk about the future we want to live in and explore the various futures we must avoid. Science fiction gives us a language so that we can have a conversation about the future. SF prototypes are tools to develop that language. The stories, movies and comics that we make can get researchers, designers, scientists, engineers, professors, politicians, philosophers and just everyday average people thinking about science in a new and creative way by using science fueled stories that capture our imaginations. SF prototypes let us imagine the future, to think through the ethical implications of technologies, play with possible benefits, explore possible tragedies and ultimately engage in a deeper conversation about science, technology and our future.

THE FUTURE IS IN YOUR HANDS

I am a futurist at the Intel Corporation. My brother says that if I were a character in a science fiction movie, when I came onto the screen and announced that I was a futurist from the Intel Corporation, pretty much everyone in the audience would assume that I was evil. That I was the villain! But I am not really that kind of futurist. My type of futurism is very pragmatic and realistic. It is my job to envision how people will use computers and intelligent devices in the future. I am writing this in 2011, it is my job to develop a practical vision for computers and gadgets in the year 2020 and beyond.

Now most of this work is not as exciting as it sounds. To have a vision for computing in 2020, you need to have a vision for what the world will actually look like in 2020. This involves a lot of research profiles, government reports, projections, market analysis, studies and a mass of data about where our world economies, environment and infrastructure are headed. The biggest let down about the year 2020 is that is looks a lot like 2010. Just like 2010 looked a lot like 2000 in many ways. But there are some bright spots, some cool new gadgets, some nearly magic technologies and some things that could change how we live, work and have fun.

Now future casting may sound like science fiction but it is not. It is actually very pragmatic. Intel makes microprocessors, Chips, the brains inside your computer. To design, build and produce a microprocessor take about 5 to 10 years. What that means is that today, in 2011, we have to have a pretty good idea of what people will want to be doing in 2017. We are building that 2017 chip today. I actually had a meeting about it this afternoon.

Future casting is not about predicting the future. Like I said before, the future is not set, we build our future everyday by the decisions we make and the things we do. Because of this, it is really impossible and useless to try and predict the future. Future casting is a process that we use to develop a vision for the future. We pull together trends, global projections and technology development into this vision and then iterate it over time. The future is always in motion, so we have to keep moving as well, adjusting to innovation and changes in culture. The key to this process is to make things, develop prototypes and generally just create stuff that expresses the future as we currently see it. Then, we share these artifacts with people to start or continue the conversation about where we are headed. Often, the things that we create are complex data models that we can analyze and discuss, but just as often, these prototypes are science fiction stories, movies and comics.

SF prototyping is an important part of future casting. We use these prototypes to envision technologies in the lives of real people all over the world. SF prototyping allows us to explore and iterate how technology can shape and be shaped by the people who use it. To date, probably the most public example of how we have used SF prototyping at Intel is The Tomorrow Project. In Germany, in 2010, we published *Uber Morgen*, roughly translated, it means "the day after tomorrow." The English language version of the work was called *The Tomorrow Project* (http://newsroom.intel.com/docs/DOC-1490).

The Tomorrow Project is a short story anthology of four works by world renowned science fiction authors and futurists. To produce the stories, we gave them access to the technology development work we are doing in Intel's labs. Our engineers showed them the work we are doing in robotics (robot assistants for interior spaces where people live or work), telematics (making road transportation more intelligent; cars, trucks, buses, street signs that could be equipped with IT systems to communicate and reduce congestion), photonics (transferring massive amounts of data using light), dynamic physical rendering (using intelligent material consisting of thousands of tiny robots to collectively facilitate telepresence, teleoperation and distributed sensing) and finally, the fascinating intelligent devices (devices equipped with sensors to analyze environmental pollution as well as a personal health and emotions, allowing people to interact with devices using gestures and even just their thoughts).

What was striking about each story was that even though they are all science fiction stories, they are all first and foremost, stories about people. Each story was unique in its own vision and portrayal of life in the future, but each of them was also extraordinarily good at capturing the human drama of the future. These stories were not about technology, they were about the complex and fascinating lives of their characters. Technology was simply a part of the drama.

In the collection, English science fiction writer Scarlett Thomas' story, *The Drop*, gives us a portrait of a family in a world that is mundane and familiar yet ingenious in its technological connections. German author Markus Heitz's *Blink of an Eye* is a fascinating cautionary tale, pitting our human wants and desires against our ability to construct a future that we may not want to live in. American Douglas Rushkoff's *Last Day of Work* tells us about Dr. Leon Spiegel's last day of work, literally the last human to work. With intelligence and foresight, Rushkoff ultimately challenges what it means to be human. And finally, England's Ray Hammond's *The Mercy Dash* gives us a couple's pulse-pounding break-neck race to save the life of a loved one. It is a race that is both helped and hindered by a complex landscape of devices, sensors and connections.

These stories ultimately show us that the stories of our future are not about technology, megatrends or predictions. They show us that the future is about people. By building these visions for the future, we can explore the innovations that Intel is building in our labs today and imagine them in real-life situations tomorrow.

But I think SF prototyping is a bit more important than that. Remember what I said before that the future is not set … that the future is made everyday by the actions of people? I truly believe this. I also believe that it is incredibly important that we are all active participants in the future. What kind of future do we want to live in? What are we afraid of ? What should our future look like all over the world? Well, if we are the ones who control our future, then we should do something about it!

Creating SF prototypes, writing stories, making movies and drawing comics about the future are one of those things that people can do to actually change the future. Think about the power of science fiction to capture people's imagination and change the way we think. *WarGames* was just one of the science fiction stories that did that for me when I was a kid. If we can capture people's imaginations and get us all talking about the future, then we can affect that future, we can build it by what we care doing today. Your future is truly in your hands.

"SHALL WE PLAY A GAME?": WHAT YOU CAN EXPECT FROM THIS BOOK

The idea of using science fiction as a prototype or development tool for the future is a pretty new idea. Certainly, science fiction and science fact have had long and fruitful relationship in the past. Science fiction stories and movies have inspired generations of scientists, engineers and developers. Similarly, scientific breakthroughs and technology innovation have revved up the imaginations of writers, filmmakers and artists.

In their 2012 book, *Make it So*, authors Nathan Shedroff and Chris Noessel delve into the relationship between interface design in science fiction and the real world. Shedroff is the program chair for the California College of Arts' MBA in Design Strategy program, and Noessel is Director of interaction design at Cooper in San Francisco. Their book is an in-depth and scholarly analysis of the technology and user interfaces (UI) which have been developed and used in science fiction film and television. They look at how scifi interfaces have changed through time and what we can learn from them.

One fun and thoughtful example can be found in how George Lucas' *Star Wars* uses holograms. The production designers' use of holographic technology (in academic circles they are more accurately called "volumetric displays") can tell us, the audience, a lot about not only the Empire and the Jedi technology but also their social order. In turn, this gives us some clues about what it might be like to work for the Empire as opposed to the Jedi. But on a more serious level, we can also learn what we should and should not do when we are designing social technologies; whether they be volumetric displays or simply a video call. If you have not seen the Star Wars movies recently, I will try to fill in a little of the story, but really you should just watch them again.

In *Star Wars Episode V: The Empire Strikes Back*, when the villain of the movie, Darth Vader, a Dark Lord of the Sith and leader of the Empire's Imperial star fleet, calls his boss and Sith Master Emperor Palpatine, the Emperor is represented by a massive, room-sized hologram of his head. We can only see the Emperors head and it looms over Vader. There is obviously a serious chain of command when you work for the Empire. It is imposing and quite scary.

In contrast to this, in *Star Wars Episode I: The Phantom Menace*, when we see the Jedi council meeting, it includes some holograms of Jedis attending remotely. Each and every hologram is their natural size. The Jedi take pains to make sure that the holograms of the council match their egalitarian principles. Powerful Jedi master, Yoda, is his usual small self, and Jedi Master Mace Windu, played by American actor Samuel L. Jackson, remains his usual six-foot three-inch height. This shows us that when you are a Jedi, you are who you are. There is no hierarchy, and all are equal.

Shedroff and Noessel's book is filled with insights for both professional designers and science fiction movie buffs. Also, in the book, the authors have a great real-world example of how science fiction and science fact have influenced each other.

In 2000, Douglas Caldwell, an engineer with the US Army Topographic Engineering Center, went to see the film X-Men by his teenaged son. He wasn't really a fan of science fiction, but while watching he saw something extraordinary that changed the direction his work. In a scene, the X-Men mutants are gathered around a 3D "pin" map big enough for all to see the representation of their plan to thwart the villain Magneto. As they describe the mission and the coordination needed, the map shifts to reveal different scales, topologies, and locations, as needed.

This isn't a unique theme in science fiction, but these kinds of collaborative map tools are usually depicted as visualizations (sometimes holographic, sometimes projected on a flat surface, and sometimes screen-based). In X-Men, however, the map is three-dimensional and rendered as solids constantly shifting (as if each solid pixel was a magnetically controlled metal cube).

What was astonishing to Mr. Caldwell was that this interface, in front of him, was a more advanced vision for what he was in charge of developing. One of the responsibilities of the US Army Topographic Engineering Center is to provide generals in the field with "sand tables," 3D representations of terrain in the theater of war that helps generals and other military personnel to coordinate military tactics. These devices are descendants of actual tables filled with sand that could be built up to mimic actual terrain. More advanced versions were needed to allow complex actions to be planned without hauling heavy and bulky equipment around the world.

When Mr. Caldwell saw the 3D table in X-Men, he instantly saw a solution he had never considered. He knew that the technology in the film was fantasy but he also surmised that the user experience was valid. At least, in this very limited snapshot, it worked for him and he tho.

What is different about SF prototypes is that they endeavor to create science fiction developed specifically on science fact as a way to inspire a conversation about the future and ultimately explore the implications of that science on the everyday lives of people. In this way, an SF prototype is a tool that can help us build better technology and sometimes practically speed up the development of hardware and software (more about that in Chapter 7).

In *WarGames*, Joshua the AI asked, "Shall we play a game?" SF prototypes are a kind of game; a thought experiment that imagines what would really happen if … What would happen if this technology truly went wrong? What would happen if everyone on the planet had access to this? What's the best thing that could happen? What are the legal, ethical and moral implications? What does this mean for our future? What kind of future do we want to live in? … and I can think of no better questions to try and answer.

• • • •

CHAPTER 2

Religious Robots and Runaway Were-Tigers: A Brief Overview of the Science and the Fiction that Went Into Two SF Prototypes

First, a little background.

The blending of science fiction and science fact is nothing new. Their symbiotic relationship stretches back in history for hundreds of years. No one would really argue that scientific research and technology inspires writers to dream up thrilling stories and amazing new worlds. Likewise, generations of scientists have had their imaginations set on fire by science fiction stories, inspiring them to devote their lives to science.

I was speaking a while back at a science convention not too long ago about the link between science fiction and science fact and SF prototypes. After my talk, at least five different roboticists pulled me aside and told me that the real reason they had gotten into robots in the first place was because of C3PO and R2D2 in George Lucas' 1977 space opera classic, *Star Wars*. When each of these incredibly intelligent scientists confessed this to me, they spoke in a hushed voice as if they were telling me a dirty little secret. But I quickly laughed and told them that they were not alone!

It is well documented that science fiction has inspired generations of scientists, researchers and even astronauts. British science fiction author, inventor and futurist, Arthur C. Clark, summed it up this way in his essay "Aspects of Science Fiction": "All of the pioneers of astronautics were inspired by Jules Verne, and several (e.g., Goddard, Oberth, von Braun) actually wrote fiction to popularize their ideas. And I know from personal experience that many American astronauts and Soviet cosmonauts were inspired to take up their careers by the space travel stories they read as children (one of my proudest possessions is a little monograph, Wingless on Luna, bearing the inscription, 'To Arthur, who visualized the nuances of lunar flying long before I experienced them!—Neil Armstrong')" (**Clark, 1999**).

Clark's story does a good job of showing that science fact and fiction have been explicitly intermingled for most of the twentieth century. Physicists and rocketry pioneers, Robert Goddard, Hermann Oberth and Wernher von Braum, used stories as a way to popularize their thinking, while astronaut Neil Armstrong was inspired and driven by Clark's writing.

The writer J.G. Ballard said, "Science Fiction is the only true literature of the twentieth century." This is a provocative and challenging statement. Is it true? Looking back, science fiction captured the wonder and promise of the magnificent world that many believed and hoped our scientific advances would bring us. With the end of WWII and the development of the atomic bomb, science fiction gave us an outlet to explore our darkest fears that may be by splitting the atom that the human race had overstepped its bounds, unleashing a demon that would eventually wipe us all off the face of the earth. After the 1960s, life seemed to imitate science fiction; we landed on the moon, robots built our cars, AI's landed out planes and ultimately the Internet changed our lives. All of this was imagined first in science fiction.

In the 1970s, an entire sub-genre of science fiction sprang up around writers that aligned themselves closely with what many people call the "hard sciences." What they mean by hard science are things like computer science, astronomy, physics and chemistry. American science fiction author, Alan Steele, defines the subgenre as "the form of imaginative literature that uses either established or carefully extrapolated science as its backbone" (**Steel, 1992**). In the chapter "Becoming an SF author" from *The Complete Idiot's Guide to Publishing Science Fiction*, the authors point out that, "Much of the wonder in hard SF comes from discovering just how wild and fantastic the natural world is. It's also fun to explore what's possible for the future of humanity, and this is what hard SF excels at. Some of the possibilities are mindboggling, but hard SF requires that we consider them as real possibilities—things like nanotechnology, genetically engineered immortality, interstellar travel and artificial intelligences that surpass our own" (**Doctorow and Schroeder, 2000**).

Many critics see hard SF as the only true science fiction because it is based on real science as opposed to pseudoscience or even science fantasy which concerns itself with notions of time travel, extrasensory powers and super heroes. Regardless of where you stand in this debate, it is clear that scientific research, theory and practice have a considerable and sometimes polarizing effect on science fiction.

The 21st century has brought us some fascinating explorations into the specific relationship between science fiction and actual technologies that are being built. Two people who have done some really interesting work in this area are Bell and Dourish. Dr. Genevieve Bell is a cultural anthropologist and head of the Interaction and Experience Lab at the Intel Corporation. Paul Dourish is as professor of Informatics in the Donald Bren School of Information and Computer Sciences at the University of California, Irvine. In 2009, these two wrote a paper called "Resistance is Futile: Reading Science Fiction and Ubiquitous Computing." In the paper, they explore how understanding science fiction, in their case popular British and American television shows like *Dr. Who* and *Star Trek*, is essential when designing new technologies.

Bell and Dourish think that science fiction can be employed as a tool for design. They think that the futuristic visions expressed in science fiction television shows can be used to understand people's collective imagining of what future technologies might look like. Looking at *Star Trek* and *Dr. Who* and understanding their effect on TV viewers would allow technology developers and software programmers to develop technologies and products that are easily understood and used by people.

Arguably, a range of contemporary technologies—from PDAs to cell phones—have adopted their forms and functions from science fiction. As in the famous case of

British science fiction author Arthur C. Clark's speculative "invention" of the communications satellite, science fiction does not merely anticipate but actively shapes technological futures through its effect on the collective imagination. At the same time, science fiction in popular culture provides a context in which new technological developments are understood. Science fiction visions appear as prototypes for technological environments." (**Dourish and Bell, 2009**)

Artist and technologist, Julian Bleecker, takes this idea and runs with it in *Design Fiction: A short essay on design, science, fact and fiction*. Bleecker thinks that using science fiction and future visions can be an incredibly productive tool for developing new technologies. He sees the interplay between science fiction and science fact as a fertile ground for the inspiration and creation of physical prototypes, conceptual inventions and actual technology. Bleecker imagines that scientists and technologists could use these prototypes to expand their thinking. Sometimes, the creations can be real functioning devices, but sometimes they can just be wildly new concepts. These prototypes become a design tool that is both real and fake, operational and symbolic, serious and ironic. Bleecker describes them as a "conflation of design, science fact and science fiction…an amalgamation of practices that together bends the expectations as to what each does on its own and ties them together into something new."

Like Bleecker's Design Fictions, SF prototypes seek a productive middle ground between fact and fiction. But when we say prototype what exactly do we mean? How are we defining prototype?

WHAT IS A PROTOTYPE?

There is a lot of debate as to what a prototype actually is. The word *prototype* has different meanings, depending upon the business or market you are talking about. A prototype in software design is wildly different than a prototype to the automotive industry. Both in academia and in the field of design, the definition is hotly debated. There is so much robust discussion and arguing for good reason.

In the technology development process, a prototype is important to the design and development process. This is because in both hardware and software development, there is great uncertainty as to whether a new design will actually do what is desired. The website *Wikipedia* explains that new designs often have unexpected problems. A prototype is often used as part of the product design process to allow engineers and designers the ability to explore design alternatives, test theories and confirm performance prior to starting production of a new product. Engineers use their experience to tailor the prototype according to the specific unknowns still present in the intended design.

But we are going to take a step back to look at prototypes a bit differently. I believe that a prototype is really just a fiction. A prototype is a story or a fictional depiction of a product. The prototype is not the actual thing that we want to build; it is an example, a rough approximation of the thing we hope to one day build. This works for software just as much as it works for concept cars. Prototypes are not the thing, they are the story or the fiction about the thing that we hope to build. We then use these fictions to get our minds around what that thing might one day be and we also use it to explain together what we hope to build.

Nathan Shedroff has an interesting take on the nature of design and prototypes. Shedroff is the program chair for the California College of Arts' MBA Strategy in Design program and is the author of several books on design like *Experience Design* and *Design Is the Problem: The Future of Design Must be Sustainable*. He believes that all design is a kind of fiction. On stage at Macworld in 2011, he said that, "Every sketch, model, and prototype is an elaborate fiction on the road to something becoming real. But, it's still fiction until (and if) it actually gets built. Think of all of the design work *left on the drawing board*. Every logo, model, variation, and failed prototype was a fiction that didn't come to be. In truth, all business plans—and planning—are fiction. Anyone who has ever created Proforma financial statements or written a business plan has confronted this. It's all made-up, even if it's based on sound assumptions."

SF prototyping, as a kind of *fictional* prototyping, provides a new lens through which emerging theories can be viewed differently, explored freely and ultimately developed further. Bleecker makes an excellent observation: "Productively confusing science fact and science fiction may be the only way for the science of fact to reach beyond itself and achieve more than incremental forms of innovation." (Bleecker, 2009) It is precisely this productive confusion and fusion of fact and fiction that can unlock, broaden and expand the boundaries of current scientific thinking.

TWO EXAMPLES OF SF PROTOTYPES

To get things started let us take a look at two SF prototypes. I have summarized both the stories and the scientific papers and research that went into them. If you want to dive in completely, you can take a look at the complete SF prototypes in Appendix A.

Religious Robots: Trouble at the Piazzi Mine

In 2008, I developed an SF prototype based upon some robotics and AI work that was going on in England and Malaysia. The basic idea of the work asked: What if robots and AI made both rational and irrational decisions? The story went like this:

There is trouble at the Piazzi mine on the dwarf planet Ceres 1. The facility's all-robotic mining crew is behaving strangely. One day a week, they are shutting themselves down completely. No work is getting done. The owners of the mine are losing 2.5 billion euros a week. Something has to be done.

Dr. Simon Egerton, a roboticist and freelance investigator, is brought in to find out what is really happening. Egerton and his body guard, Nigel Kempwright, are given two days to uncover the mystery.

"What is all this nonsense about mining bots going to church on Sundays?" Kempwright tugged on his shaggy blond hair. "Are they serious? I mean really.

Are they serious? Who's the bright guy who can't just reprogram them or something?"

"They tried that I think," Egerton replied. "It didn't work. I guess everything they've tried hasn't worked."

Landing at the Piazzi Mine … was like descending into the mouth of hell … the place looked like a giant worm had burrowed deep into the rock, leaving its eggs along the interior of the shaft to hatch and continue its work. These eggs were the buildings and outbuildings of the mining facility, screwed precariously into the sides of the tunnel, and connected by a maze of spidery catwalks and walkways. Two massive mining chutes dove down into the blackness of the pit.

Egerton and Kempwright emerged from the battered shuttle to the vast loading deck of the Piazzi Mine … The deck was crowded with the tremendous robotic loaders, recklessly filling mega-haulers. Their shuttle was dwarfed by the size of the operation and Egerton and Kempwright were barely visible amongst the giants.

Everything looked normal. Egerton imagined the AI agent sub-systems orchestrating every action of the various bots and machines; reacting to the mine's changing conditions, updating, correcting, always taking in information and reacting. When Egerton thought of the dirty little mine in that way it was quite delicate and beautiful; hardware and software dancing elegantly together.

"This is our only hotel at the mine," the security bot said stopping in front of a small shabby trailer. "It doesn't get much use as you can imagine."

The automatic doors hissed open and Egerton could see another bot waiting for them inside.

"And what's that?" Kempwright asked, still standing on the narrow street. He pointed to a massive warehouse directly opposite the mine's main shaft.

Egerton stepped back out into the street and instantly saw the massive structure. It stood out not because of its size but because it was spotless in a swarm of grime.

"That is the church," the security bot replied calmly.

The church was larger than he had expected. The calm air smelled of industrial lubricant and electricity. Across the expanse hung a fifty foot electric-yellow cross, crudely constructed out of two mine support beams. Lined up in front of the cross, in massive neat rows was every bot from the Piazzi mine.

Moving down the center aisle, Egerton could see it wasn't just the ambulatory bots, the ones that moved freely that had come. It was all the robots. Even the massive diggers had been un-bolted from the mine shafts and carried in; the wheel-less, the legless, the immobile; no bot had been left behind. Egerton found himself searching the air for signs of the unseen nano-bots.

Stopping mid-way up the aisle, Egerton knew he had to get back to catch the shuttle. Easing back to the door with quiet *reverent steps, he searched the bots for any sign of activity. Their optical scanners were closed, their activity lights dimmed. Nothing moved, nothing stirred. It was so still that Egerton could hear the blood coursing through his ears.*

* * *

"Well, then I guess we aren't understanding each other then because it sounds like you're tell me that our mining bots now somehow believe in God." XienCheng's [the mine's administrator] jowls shook with frustration.

"It's not that," Egerton tried again. "It's not God. They don't believe in God. They believe in going to church, in the action of going to church."

"But that doesn't make any sense," XienCheng interrupted.

"Yes!" Egerton slammed his hand down on the table. "Now you're getting it."

XienCheng drew back in an awkward fear. "No. No, I'm not."

"They aren't doing anything in the church," Egerton explained. "They aren't worshiping God or saying prayers or holding a service; they're just going to church. It's the action and it's not supposed to make sense. That's the point. For some reason they latched onto the idea but now they need it. It keeps them safe." (Johnson, 2009)

Mystery solved! The robots at the Piazzi Mine were going to church precisely because it was irrational. Like some human behavior, irrational actions can be as constructive as rational decisions. Most people think that robots and artificial intelligences (AI) can only make rational decisions. But we humans make good decisions and bad decisions, and it is because of this that we learn much faster than if we only made "correct" or rational decisions. It is this complexity of learning that allows humans to operate in highly complex environments.

Now you might be saying: "But that's just science fiction!"

But you would be wrong …

The following abstract is from a paper with the impressive title: *Using Multiple Personas in Service Robots to Improve Exploration Strategies When Mapping New Environments.* This research was the work of three scientists Dr. Simon Egerton, Dr. Victor Callaghan, and Dr. Graham. [*Note: You may have noticed that the name of the main character in Nebulous Mechanisms and one of the authors of the research paper are the same: Dr. Simon Egerton. This was originally done as a homage to Dr. Egerton as he was the first person to explain to me the work that the team was doing over a pint of beer in Seattle Washington. When I named the character none of us realized that the SF prototypes would be so popular and effective, spawning a whole series stories. More about that in Chapter 7. If it is any conciliation—the main character of the stories pronounces his name "Egg-er-ton" whereas the good doctor pronounces his name "Edge-er-ton." I am indebted to Dr. Egerton ("Edge-er-ton") not only for is mind-expanding research but also for being so understanding and having an excellent sense of humor about me dragging his name all over the world.*] Published in 2008, the paper explains the emerging research by the scientists who were using a concept from philosophy and psychoanalysis in their robotics and AI development work.

Abstract

In this paper we describe our work on geometrical and perception models for service robots that will support people living in future Digital Homes. We present a model that captures descriptions of both the physical and perceptual environment space. We present a summary of experimental results that show that these models are well suited to support service robot navigation in complex domestic worlds, such as digital homes.

Finally, by way of introducing some of our current, but unpublished, research we present some ideas from philosophy and psychoanalytic studies which we use to speculate on the possibility of extending this model to include representations of persistent experiences in the form of multiple personas which we hypothesize might be applied to improve the performance of services robots by providing a mechanism to vary the balance of current and past experiences in control decisions which appear to serve people well. (**Egerton, Callaghan and Clarke, 2008**)

The idea of multiple personalities is not a new concept to people. The concept was brought to the attention of popular culture with the 1973 book *Sybil* by Flora Rheta Schreiber. The book was about a patient, Shirley Ardell Mason, who was being treated for dissociative identity disorder by her about her psychoanalyst Cornelia B. Wilbur. Mason's condition was more popularly known as multiple personality disorder. In 1976, a popular made-for-television mini-series was produced based on the book starring Sally Field as the patient and Joanne Woodward in the role of Sybil's psychiatrist. Field won an Emmy award for her performance. The Internet Movie Database describes the movie like this: "The true story of a young woman named Sybil, whose childhood was so harrowing to her that she developed at least 13 different personalities" (**IMDB, 2010**).

Egerton, Callaghan, and Clarke take the multiple personalities and apply it to the development of robots and AI. They expand multiple *personalities* into multiple *personas*, which has a little bit of an expanded meaning. Most think of a persona as the role that we as people play in society and in our lives, whereas a personality is a collection of traits that make up a person. The highly influential Swiss psychiatrist Carl Jung defines personas as "the mask or appearance one presents to the world." Egerton, Callaghan, and Clarke use this idea to radically rethink the design of AI and software systems.

> *Multiple personas in philosophy and psychoanalysis are seen as one explanation for irrationality and are based upon the generally pathological process of splitting. In this instance splitting for functionally sound reasons is suggested as a possible aide to robust and efficient working within a variety of different contexts. In this sense it is more akin to the benign forms of splitting that allow us to be parents, siblings, children, workers, partners etc.* (**Egerton, Callaghan and Clarke, 2008**)

The scientists are using "multiple personas" in a constructive and pragmatic way in the development of their robot's AI. The splitting of the personalities in this case is not negative as with Sybil. They take a very different approach. They view the splitting of the *personalities* into *personas* as a highly efficient and functional way to segment a robot's intelligence or brain. By separating the functions of the robot into these personas, it becomes easier for the scientists to design and develop the robot's intelligence.

> *This persona based approach to the architecture will allow us to explore the hypothesis that multiple personas guide our actions, that we do not make decisions purely on our immediate sensing of the world. By having a reservoir of specialised personas to call upon, the persistent and evolving nature of such personas would allow us to explore the value of accumulated experience that in us manifests itself as a somewhat ill defined 'self'; which, when making decisions, occasionally overrides the logical nature of the world, akin to what might appear to be irrationality, putting it down to nebulous mechanisms such as "a hunch" or "a feeling." This approach will allow us to open up a line of research to explore the nature and value of such abstracted*

personas and their dynamics. (**Egerton, Callaghan and Clarke, 2008**)

Nebulous Mechanisms acts as an SF prototype for the scientists' vision for an approach to programming AI. The story allows us to imagine some extreme effects of their approach and what it might mean to have "irrational" robots. What could be the positive and negative effects of this? Ultimately, *Nebulous Mechanisms* pushed us to think about Egerton, Callaghan, and Clarke's work in new ways and from different angles. The result of this new perspective is a now expanded concept of the scientists' original work.

In Chapter 7, we will explore other examples of SF prototypes in the Dr. Simon Egerton Stories and how researchers and scientists are using the stories to expand and further define their development of software, AI, and robots.

Runaway Were-Tigers

On July 19, 2010, the 1st International Workshop on "Creative Science—Science Fiction Prototyping for Research" (CS'10) was held in Kuala Lumpur, Malaysia. This workshop explored the use of science fiction as a means to motivate and direct research into new technologies and consumer products.

The workshop did this by creating science fiction stories grounded in current science and engineering research that are written for the explicit purpose of acting as prototypes for people to explore a wide variety of futures. These "prototypes" were created by scientists and engineers to stretch their work or by, for example, writers, school children and members of the public to influence the work of researchers. The outcome of these interactions was then to be fed back to guide further research and development activities.

The website for the workshop describes its approach to SF prototypes in this way: SF prototypes involve the widest section of the population in determining the science research agenda, thereby making science investment and science output more useful to everyone ranging from companies, through scientists and engineers to the public, consumers and the government that indirectly fund R&D. In this way, fictional prototypes provide a powerful interdisciplinary tool to enhance the traditional practices of research, design and market research. The goals of the workshop were to act as a catalyst of this new approach by acting as a forum where researchers from differing disciplines (notably science fact and science fiction) can come together to explore how to develop this area (**Creative Science Foundation, 2010**).

One of the SF prototypes that was presented at the workshop was called *The Were-Tigers of Belum* by Kar-Seng Loke and Simon Egerton. I have compiled an overview below. The full text of the story can be found in Appendix A.

The screen flashed. Raja jumped back to his seat. He selected the hotspot on the screen to get the readout on it.

Just about then, the sun in the between the hills began to show itself, casting longish shadows on this nondescript shop lot in the leafy suburb neighbourhood of TTDI, in Kuala Lumpur. The plain signage on the door simply read A.E.O.N....

* * *

Just then, Kim let out a whistle, and called out to Raja, "Look at this model animation of the gait, Raja, they look weird, certainly not a normal gait of a tiger."

"Oh my god, I don't believe this." Raja gasped. Then he continued, "You know, Kim, the natives believe that a tiger-spirit roams the forest. They believed that the tigers are the guardians of the forest, and its protector. These spirits are the re-manifestation of their ancestral spirits. To see them is an occurrence of dire portent, a harbinger. It is also believed that these spirits can be called into existence by the bomohs, or by some momentous future event … usually a sign of something that is going to happen, be it good or bad."

"Surely you don't believe in that stuff, anyway, what has it to do with this?"

"Well what if it's true, there might be something to it? What if we have triggered something in the deep forest, our machines have penetrated into parts of the forest where nobody has ever gone before, what if we have violated the sanctity of the tigers most ancestral place and awakened the semangat of the forest!?"

"Aw c'mon Raja, stop pulling my leg, this is the 21st century, you are a good scientist and you are still talking about *spirits? What have been smoking lately?"*

"What does this century have to do with anything? Don't you see all this datuk-datuk by the roadside everywhere, with all their elaborate shrines and offerings? Rites are still being performed prior to important events."

"And … your point?" asked Kim.

"Look, you said those are not tiger-like gaits, yet the software has identified them as a tiger like gait, it is a tiger that does not have a tiger like gait. We know our software and AI works, we tested it, rigorously. So, what does that tell you? Do you know that it is believed that the bomoh with sufficient ilmu can transform into were-tigers. These were-tigers can be recognized by the lack of the groove in the upper-lip and by their gait. This is because their heels are reversed! Don't you see, this is precisely what the software has detected, the unusual gait is caused by the reversed heels, can't you see it in the sensor pattern on the screen?"

"Are you seriously suggesting that we are tracking were-tigers?" asked Kim, incredulously.

"Well …" Raja, confused, stalled as he couldn't really put in a satisfyingly coherent reply. (**Loke and Egerton, 2010**)

This too is not science fiction. The authors of the story have been doing work on a sensor-based environmental monitoring system. In the introduction to the story, they explained their approach:

The environment and the world we inhabit today is perhaps the most precious gift we have to pass onto the next generation. Those who will inherit tomorrows' environment and tomorrows' world will no doubt question how we managed their legacy. To help us understand the complexities and sensitivities of our finely interwoven eco system and our effects on that system, we need to build accurate models from which we can derive theory, make predictions and define policy. A complete model would measure all forms of environmental data, both flora and fauna, such as plants, animals and micro bacteria, across the world, measured at frequent intervals, ideally in real-time [1]. However ideal this maybe, it is currently very impractical, there are too many species to measure and monitor, and data collection if often tedious and time-consuming and on the whole, carried out less frequently than desired.

Since it is impractical to consider all biotic taxa for measurement, ecologists have identified a small number of key indicator species, namely, Plants (Trees), Bats, Birds, Aquatic Macro Invertebrates, Moths, Ants, Figs & Frugivores, Dung Beetles, Stingless Bees and Large Mammals, ordered for their importance as a general environmental indicator [2]. Their sensitivity and stabilities to environmental conditions such as air pollution, climatic variation, foliage-densities and so on make them a practical bio-indicator, moreover, they are present, in some combination, across all continents and environmental conditions. This commonality has the advantage of facilitating a common frame of reference for data analysis.

Data collection typically involves a protracted manual process; a good example is the collection of moth data. The collection of moth data requires the ecologist to physically travel to the area of interest, assemble the collection apparatus (light-trap(s) in this case) either camp overnight, especially if the area is in a remote location, or leave and return at a later point, the raw data need sifting and cataloguing by an expert taxonomist, picking out the targeted moth species from the other collected moths and *insects, only after this process can the processed results be used for modelling purposes [3]. This process typifies bio-indicator data collection and is the process our proposed system is designed to automate.*

In this paper [SF prototype], we have envisioned a global real-time sensor network for the automated collection of key bio-indicator data, our so named Automated Eye on Nature (AEON). Although AEON is primarily to collect biotic taxa data to model the biodiversity and health of the environment, the data could also be used to drive real-time environmental models to help us better understand the complexities of our ecosystem. Our fictional prototype explored one such extended usage, where the taxa data from large mammals, tigers in the story, was used to drive AI gait models which in turn enabled identification and behavioural tracking of tigers. This type of tracking is being actively researched, although using more conventional methods (IBID).

The authors use their SF prototype to imagine the effect their monitoring system might have on the people society and ecosystem that it is monitoring. The virtual world they have created in the SF prototype allowed them to further innovate and imagine expanded uses of the technology.

SF prototypes allow us to create multiple worlds and a wide variety of futures so that we may study and explore the intricacies of modern science. They are a powerful tool meant to enhance the traditional practices of research and design. The discoveries that we make with these prototypes can be used to question and explore current thinking on a level we have not approached in the past, namely using multiple futures and realities to test the implications and intricacies of theory. Additionally, the output of the science fiction prototype can feed information back into the science and technology development process (more info in Chapter 7), investigating and shaping how a user might encounter, explore and ultimately use that technology.

Science fiction allows us to see ourselves in a new light, in the light of a new future, one that is not our own but reflects directly upon who we are and where we might be headed. The SF prototype brings this same lens to science fact, allows us to see the multiple futures in the theory we are constructing today.

• • • •

CHAPTER 3

How to Build Your Own SF Prototype in Five Steps or Less

So now it is time to create your own SF prototype. As we talked about before, the goal of the SF prototype is to use science fact as the basis of your vision of the future. In this chapter, we have broken down the SF prototyping process into five discreet steps. Each step in the process will take you through a kind of framework for getting started and collaborating. At the end of the five steps, you will have an outline that you can use to expand, develop and build your own SF prototype.

The final form of the SF prototype is up to you. Later, in this book, we will give you some examples of the three different forms of SF prototypes: short stories, movies and comic books. Each chapter on the different forms will give you a little history and background along with a discussion with an industry expert. These experts will provide you with their own personal take on the form of storytelling and how to best use the SF prototype. At the end of each chapter, there are also some practical steps to take for your outline and build it into an engaging prototype.

But, first, we have to pull together an outline.

THE OUTLINE

The outline of the SF prototype is where your ideas and reflections on the broader contextual issues will really get explored. The outline will force you to think about science in a realistic setting of people and society, without forcing you to actually become a science fiction writer (although there will be nonrequired opportunities to flesh your outlines into complete stories, if you so desire).

The purpose of the outline is to capture the *idea* behind the story and put it into a *plot*. Alan Moore, the legendary comic book writer and creator of *The Watchmen, V for Vendetta* and the *Sandman* series, describes the distinction between the *idea* and the *plot* in this way:

> *The idea is what is the story is about; not the plot of the story, or the unfolding of the events within the story, but what the story is essentially about. As an example from my own work (not because it's a particularly good example but because I can speak about the work with more authority about it than I can the work of other people). I would cite issue #40 of Swamp Thing, "The Curse."*
>
> *The story was about the difficulties endured by women in masculine societies; using the common taboo of menstruation as a central motif. This was not the plot of the story—the plot concerned a young married woman moving into a new home built upon the site of an old Indian lodge and finding herself possessed by the dominating spirit that still resided there, turning her into a form of a werewolf.* (**Moore, 2008**)

Moore gives us a great way of looking at the difference between the *idea* and the *plot*. When we start to think about constructing our science fiction prototype, the *idea* of the story will be your topic, the scientific issue that you draw from papers and experiments. The *plot* of the story is what you will contract from your outline.

Think back to our previous example in Chapter 1. In the story *Nebulous Mechanisms*, the *idea* of the story comes from the paper *Using Multiple Personas in Service Robots to Improve Exploration Strategies when Mapping New Environments*. The paper explores the benefits of building irrationality into the artificial intelligence of domestic robots to improve their ability to adapt to complex environments. The *plot* of the story revolves around Dr. Simon

Egerton's investigation of why the robots from the Ceres mine have started going to church on Sundays.

In *Nebulous Mechanisms*, the *idea* is why the story is being told, it is the idea and the theory that are being worked out in the fiction. The *plot* is what actually happens in the narrative. It is a linear set of events involving characters, locations and situations where we can explore the implications of the idea. We can put the idea into a real-world setting and see how it plays out and better understand the idea's effect on both the characters and the locations.

Dean R. Koontz is a powerhouse writer. He has been on the best selling fiction lists for over 30 years, and 24 of his titles have reached the number one spot. So it is pretty easy to say that Koontz knows quite a lot about how to put together a story. What many people do not know is that back in 1981, Koontz wrote a book on writing called: *How to Write Best Selling Fiction*. It is a very practical book that discusses Koontz's ideas on writing, story construction and the professional literary marketplace.

In his book, Dean Koontz describes science fiction plots (he calls them category fiction or genre fiction) as being a little different than other kinds of writing.

> *The plot is usually the skeleton and the tendons and the vital organs and the muscle ... a strong plot—one that is based on an ever-worsening series of complications—is essential.* (Koontz, 1981)

For your SF prototype, you will outline the plot and explore the implications of your topic. A story outline is as Koontz describes it, *the skeleton*, of the story. The outline provides a step-by-step description of what happens in the story. In most cases, the outline is not written in prose like a story. For our purposes, a list of events and description will serve to describe the action in your fictional world. To help things along, I have provided a rough structure for your outline below (Figure 1).

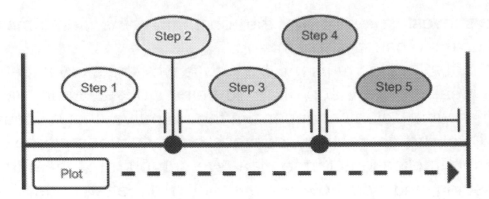

FIGURE 1: The five-step process.

THE FIVE STEPS

The following five steps break down the SF prototyping process into sequential accomplishable steps. They are:

Step 1: Pick Your Science and Build Your World

First, you should pick the technology, science or issue you want to explore with the prototype. Next, set up the world of your story and introduce us to the people and locations. You can answer very simple questions like who are the main characters and where will the action take place. You will also want to begin to explore an explanation of the technology in your topic.

Step 2: The Scientific Inflection Point

The introduction of the "science" or technology you are looking to explore in the prototype.

Step 3: Ramifications of the Science on People

Explore the implications and ramifications of your science on the world you have described in Step 1. What effect does the technology have? How does it change people lives? Does it create a new danger? What needs to be done to fix the problem?

Step 4: The Human Inflection Point

What did we learn from seeing the technology placed into a realistic setting? What is needed to happen to fix the problem? Does the technology need to be modified? Is there a new area for experimentation or research?

Step 5: What Did We Learn?
Explore the possible implications, solution or lessons learned from Step 4 and its human implications.

Step 1: Pick Your Science and Build Your World

Step 1 is probably the most important and most time consuming of all the steps. It is here that you will pick your topic and you will build the world where you will place your SF prototype.

The first thing to do in the process is to figure out what piece of science or technology you want to explore. This can be taken from any number of places. Many universities and teams already have research and technology they would like to expand. In Chapter 2, the story *The Were-Tigers of Belum* was written by the scientists who were actually doing the development of the sensor network for environmental monitoring. They developed their SF prototype as a tool to expand their thinking about how the sensor network could be used and what issues they might face as the technology was deployed. They also had some fun bringing back the mythical tigers as a reminder of the environmental issues that are currently affecting Malaysia.

If you are not currently working on a research and development project, you can review magazines and journals for your particular area of interest. It is usually more exciting to pick a piece of emerging research or science because the implications and effects of it are most likely not widely understood. Another more general place to look are popular science magazines like *Scientific American, New Scientist, Nature*, and *Popular Science*.

The goal is to pick a topic that grabs your imagination and gets you thinking about what *might* happen when people start using it.

When planning your SF prototype, you should begin by considering future versions of the technology you have selected to explore from you topic. You can begin by asking yourself some basic and entertaining questions:

- What are the implications of the mass adoption of the technology?
- What is the worst thing that could go wrong and how would it affect the people and locations in the story?
- What is the best thing that could happen and how would it better the lives of the people and locations of the story?
- If this technology was in an average home, how would it actually work?

Once you have started getting some ideas from these questions, you can begin to brainstorm one or more potential broader contextual issues raised by the technology in question. As you imagine the plot of your story, it is important to remember that you are placing your topic or idea in a real world. Now, granted we are talking about science fiction or your real world might be far into the future, but regardless, the world must feel real. It is still governed by the laws and logic of science. It is also important to remember that this world that you are creating needs to be populated by real people. These real people will have real problems that have nothing to do with your topic. In the future, people will still not want to go to a boring job. In the future, people will still fall in love and some will have their hearts broken. In the future, we still will feel too lazy to take out the trash.

The setting for most science fiction prototypes needs to be the near future. As we extrapolate out the scientific topics, the goal is to place them in a world that we know and one that will be useful to study and explore the effects of the technology. Creating a realistic background for the near future is essential.

Even a best seller like Dean R. Koontz recognized that creating a plausible near future was not an easy task.

Because most science fiction takes place in the future, the backgrounds are largely products of the writers' imaginations. The future can be researched only to a limited extent, for when it comes to saying exactly what the years ahead hold for us, even the most well-informed scientists can offer only conjecture. The SF (science fiction) writer's vision of the future must be detailed and believable, or ultimately the reader will not believe anything about the story—not the characters, the motivation, or the plot.

The near future. Structuring a story background of near future—twenty, thirty, or forty years from now—is in some way more difficult than creating an entire alien planet in some impossibly distant age, for the near-future background cannot be wholly a product of the imagination. The writer must conduct extensive research to discover what engineer and scientists project for every aspect of future life. From that data, the author then extrapolates a possible world of tomorrow, one which might logically rise out of the base of the future which we are building today. (Koontz, 1981)

This is good news! As Koontz points out, it is important that the writer of a near-future story understands what engineers and scientists are working on and projecting for the future. In the case of science fiction prototypes, much of this is accomplished when the topic has been selected. The fictional near-future world of the story/prototype is dictated by the actual science of the topic.

Step 2: The Scientific Inflection Point

Now that you have picked your science and you have your idea, it is time to see what happens when that science (or technology or topic) is placed into your world.

When doing this, it is important to focus specifically on the people and systems in your world. The inflection point is not about the science itself (you should explore that in Step one). Step 2 is all about the effect that this new science or technology might have on the daily lives, governments and systems in your story.

Step 3: Ramifications of the Science on People

Once you have unleashed you idea into your world, it will have knock on effects. The people in your story will adapt and change because of the technology you have introduced. You have to ask yourself: has it made their lives better? Has it made their lives worse? Have they adapted to the problems or opportunities it has brought about?

This typically is where the plot of your story, movie or comic book can get really interesting. In the following three example chapters about short stories, movies and comic books, we will give you examples of how you might have the situation or plot of your SF prototype get so bad that the people in your story must take action.

This is not only good story telling but it is also beneficial to the development of the technology or science you are exploring. Pushing your plot to the extremes (either good or bad) will expose new areas for investigation or exploration in the real world. Once you have explored and mapped the outer edges and extreme scenarios, then you can map back to the middle to a more normal and realistic situation. But going to these extremes will help give clarity to your idea.

Step 4: The Human Inflection Point

The situation in your SF prototype has gotten dire. You have gone to the extreme! The characters are at their wits end. Their lives may even be in danger. Ask yourself: what do your characters need to do to survive? What are the human ramifications of the science you have selected and the situation it has brought out?

This is the point where we learn what your characters will do to either adapt themselves to the science or technology you have introduced in Step 2 or the people in your story will alter or change the science to suit themselves. Both of these outcomes need to be in keeping with the world you have created and the science needs to stay logical as well. It is important in Step 4 that the changes be believable for the virtual world and stay within the constraints of science. If you constrain yourself to these boundaries, then the outcome of your SF prototype will be more valuable and applicable to further investigation.

Step 5: What Did We Learn?

After Step 4 and your human inflection point you must ask yourself, "What did we learn?" What have you learned from placing your science into the world you created (Step 1). This science had an effect on your world and the people, society and systems in that world (Step 2). You explored the effect on the world (Step 3) and ultimately the people in your world reacted to the changes (Step 4).

Step 5 gives you the space to explore each of the previous steps. How has your world changed? How have the people, society and systems changed? What could be done differently? What cautions do you need to pay attention to? What fears were unfounded?

Step 5 gives you the space to explore what is next with the science. What should be different? What would you improve? What must stay the same? What ramifications have you uncovered by using the SF prototyping process and how has it changed your outlook? How could it improve your research? (In Chapter 7 we explore in detail an example of what can happen when Step 5 expands the science it was based on and actually further the development of the technology.)

The five steps of the SF prototyping process are just a suggestion and a guide. They are in place to walk you through some simple steps to examine and reexamine both the science and impact that science will have. As with many processes, once you feel comfortable with it you can make deviations and your own modifications. But if you follow these five easy steps you will come away with a solid outline for a science fiction story based specifically on science fact. You can then turn this outline into whatever form you think best suits your ideas and imagination. Remember, the ultimate goal of the SF prototype process is collaboration, iteration and fueling your imagination to look at science and technology in new and exciting ways.

WRITING THE OUTLINE IN FIVE EASY STEPS: AN EXAMPLE OF *NEBULOUS MECHANISMS*

To illustrate this process, let us use an example. As we discussed above, in the story *Nebulous Mechanisms* (Johnson), the *idea* of the story comes from the paper *Using Multiple Personas in Service Robots to Improve Exploration Strategies When Mapping New Environments* (Egerton, Callaghan, and Clarke). The paper explores the benefits of building irrationality into the artificial intelligence of domestic robots to improve their ability to adapt to complex environments. The plot of the story revolves around Dr Simon Egerton's investigation of why the robots from the Ceres mine have started going to church on Sundays.

Here is the outline:

Step 1: Pick Your Science and Build Your World

- We chose the paper: *Using Multiple Personas in Service Robots to Improve Exploration Strategies when Mapping new Environments* by Egerton, Callaghan, and Clarke.
- Dr. Simon Egerton, a university researcher and roboticist, is called into a meeting on a near Earth space station for a mysterious job.
- XienCheng, a mid-level administrator, tells Egerton that there is trouble at their asteroid mine Ceres 1. They are losing money, and they want Egerton to figure out why it is happening.
- Egerton learns that the robots are going to church.

Step 2: The Scientific Inflection Point

- The robots are going to church on Sunday and no one knows why.

Step 3: Ramifications of the Science on People

- Egerton departs for the mine with a bodyguard, Kempwright. They only have two days to figure out the mystery.
- At the mine, the men discover that all of the humans are gone and that the mine is running normal.
- At the end of their tour of the mine, their menacing robot guide explains that the frighteningly clean building in the distance is the robot's *church*.
- That night Egerton is awakened by Sue Kenyon, the last human employee of the mine.
- Kenyon tells Egerton that her ex-partner was killed by the bots when he went snooping around and that Kempwright has disappeared.
- Kenyon must get drunk and pass out because the nano-bots in her pace maker will not work on Sunday, and if she is not passed out, she will die.
- Egerton wakes up to find Kempwright murdered by the bots (crucified) in his room.
- Egerton races to the shuttle pick up to escape any further violence but finds the mine deserted and all the robots in the church.
- Egerton investigates the church.

Step 4: The Human Inflection Point

- In the church, the robots are not holding service or praying or doing anything at all. They are simply going to church. It is an irrational behavior they have adopted to help them cope with the mine's complex environment and dangerous conditions.

Step 5: What Did We learn?

- Egerton returns to the near Earth satellite to deliver his report to XienCheng.
- The irrational behavior of the bots allows them to cope with the complex environment of the mine. It is precisely because it is irrational and does not make logical sense that they have adopted it. They now need it to operate.
- Egerton tells Xien Cheng that he needs to get used to this irrational behavior, the robots on the company's moon base are building an amusement park—engaging in another "irrational" behavior—riding roller coasters.

One example result from the *Nebulous Mechanisms* science fiction prototype is that although irrationality may allow artificial intelligence and domestic robots to adapt to complex environments, it is also important to bound this irrationality so that it does not harm humans (as with the Sue Kenyon example) or decrease productivity (church and roller coasters). We also ask new questions about the science:

- Can we put bounds on irrationality?
- Are there different levels of irrationality that are appropriate for different systems and robots?
- Can we build safety mechanisms into AIs and systems to safeguard from unexpected consequences of more complex computing environments?

WHAT IF …
The five steps of the SF prototyping process are merely a guide to get you started. We have found it is good to build your outline with

another research partner, colleague or collaborator. Part of the fun and excitement of the SF prototyping process is the collaboration and conversations that you will have. It all starts with science and the question: What if …

. . . .

CHAPTER 4

I, Robot: From Asimov to Doctorow: Exploring Short Fiction as an SF Prototype and a Conversation With Cory Doctorow

As you go about building your SF prototype, it might be helpful to get to know a little history and different examples of the process. First, we are going to start with fiction, the written word, as a form that your SF prototype could take. Ultimately, we are going to use the short story as one type of SF prototype that you can build from your outline. In this chapter, we are going to look into some history of the science fiction story, with examples of different authors who have married science fact with science fiction in interesting ways. We are also going to talk with activist and science fiction writer, Cory Doctorow, to get his thoughts on the genre, the relationship of science in his own writing and the SF prototype process in general.

But speaking of history, it is always good to start with the first one. Let's take a look at what most people think as the first work of science fiction ever. *Frankenstein*.

THE TEENAGER AND HER MONSTER

The legend goes like this: A small group of friends, teenagers and twenty-somethings, sit around one night in June during a thunderstorm. With nothing to do, they try to scare each other by reading aloud *Tales of the Dead*, a book of German ghost stories. Later that night, when they all go to bed, one of the girls had a frightening dream. She wrote that she saw "the hideous phantasm of a man stretched out, and then, on the working of some powerful machine, show signs of life, and stir with uneasy, half vital motion" (Shelley, 1831).

The girl's name was Mary Shelley, and the novel that grew out of the nightmare was *Frankenstein*. The other story tellers that night were her husband, the English poet, Percy Shelley; their infamous friend, Lord Byron; Byron's mistress, Claire Clairmont; and their doctor, John Polodori. Within 8 years, both her husband, Percy, and friend, Lord Byron, would be dead; Percy in a boating accident in 1822, and Byron would die fighting against the Ottoman Empire in the Greek War of Independence in 1824. Mary Shelley never remarried and supported herself by writing until she died at the age of 53 years.

Mary Shelley's gothic story is considered by most to be the first science fiction novel. *Frankenstein* is considered the first not only because of its content but also because of the time it was written and how it was heavily influenced by the science and society of its time.

England in the early 19th century was in the midst of a radical change. The industrial revolution had changed how people worked and the entire make up of English society. Science fiction legend, Brian Aldiss, along with his coauthor, David Wingrove, describes Mary Shelley and her friends' understanding of the rapidly changing world around them like this in their history of science fiction, *Trillion Year Spree*. "The Byron–Shelley circle understood themselves to be living in a new age. They felt themselves to be moderns. The study of gasses was advanced, much was determined about the composition of the atmosphere, that lightening and electricity were one and the same was already clear. Mary lived in a thoroughly Newtonian world, in which natural explorations could be sought for natural phenomena" (**Aldiss and Wingrove, 2001**).

Shelley knew she was living in a new age. An age where logic and the scientific method were reshaping how people thought of themselves and the world around them. Frankenstein takes these themes head on; exploring what it means to be human, the boundaries of science and the ethics of generating or taking life. What is a scientist's responsibility to her/his inventions? These are ethical and moral problems that are still being debated today.

Brian Aldiss also put forward that what makes *Frankenstein* so unique is that it portrays "an inescapably new perception of mankind's capabilities." Shelley was using her fiction to explore how her society and how humanity was being transformed by science and the scientific method. "In combining social criticism with new scientific ideas, Mary Shelley anticipates the methods of H.G. Wells —and of many who followed in Wells' footsteps (**Aldiss and Wingrove, 2001**).

WHEN SCIENCE CAME TO SCIENCE FICTION

The English author H.G. Wells was born in 1866 at the end of the 19th century. Dying in 1946, Wells life, experiences and writings served as a bridge to the 20th century. He is most popularly known for his novels *The Time Machine* (1895), *The Invisible Man* (1897) and *War of the Worlds* (1898). These books however were written quickly, and all of Well's science fiction was completed over a 5-year period. Much of Wells' career was spent writing contemporary novels, as well as books on history, politics and social commentary. Late in his life, Wells did not look back fondly on his most popular works. He saw them as frivolous and merely entertainments. He wrote in the preface to a collection of these stories, "It occurred to me that instead of the usual interview with the devil or magician, an ingenious use of scientific patter might with advantage be substituted … I simply brought the fetish stuff up to date, and made it as near to actual theory as possible" (**Wells, 1933**).

Wells was not so much concerned with science as he was with his own society of the day. It was not until the beginning of the 20th century that the scientific method came to science fiction.

American science fiction writer, editor and luminary, Huge Gernsback, was one of the first to overtly link the world of science fiction and the world of science fact. In the pages of his early 20th Century magazine, *The Electric Experimenter*, Gernsback placed science fiction stories next to articles about science fact. Gernsback believed that "A real electrical experimenter worthy of the name must have imagination and a vision for the future" (**Ashley, 2000**). Gernsback was such a visionary author and editor that he has been honored by the Science Fiction community. The Hugo Award, named after Gernsback, is one of the highest honors that can be bestowed upon a science fiction author.

Nearing the middle of the 20th Century, Gernsback's tradition of combining science face and science fiction continued when editor John W. Campbell encouraged a then relatively unknown science fiction author named Isaac Asimov to incorporate logic and the scientific method into his stories.

This proved to be a clear turning point from all of the science fiction that had come before. Mary Shelley's *Frankenstein* is often cited as the classic example of this kind of story. Shelley's gripping novel is truly pioneering in its subject matter and power, but the story is more metaphysical than scientific. The novel wrestles with the themes of creation and the science of man that has lost control of its creation. However, Shelley does not concern her narrative with the details of the monster's creation. In fact, Dr. Frankenstein's lab, reimagined countless times in movies and comic books, is barely described by Shelley at all in any detail or scientific way. "In a solitary chamber, or rather cell, at the top of the house, and separated from all the other apartments by a gallery and a staircase, I kept my workshop of filthy creation … the dissecting room and the slaughterhouse furnished many of my materials" (**Shelley, 1831**).

With all of these said, Shelley does not delve into the science of her story because the exploration of science is not her end goal. It is a matter of intent. Frankenstein is a masterpiece about the responsibility of science and ultimately what it means to be human. Shelley wrote that she wanted her story to "speak to the mysterious fears of our nature" (**Aldiss and Wingrove, 2001**). It is not about the nuts and bolts of how one might create human life by sewing together the corpses of criminals and using lightening to give them consciousness.

Editor Campbell and writer Asimov set about to create a new kind of science fiction story—A story that was based upon logic and the scientific method. From this idea came Asimov's hugely popular robot stories and the three laws of robotic that fueled the logic behind them.

Robbie, the first robot story, appeared in 1939, although it was based on this idea of using an underlying logic to fuel the narrative, Asimov's three laws were not fully formed until 1942 and his story *Runaround*. The laws are as follows:

1. A robot may not injure a human being or, through inaction, allow a human being to come to harm.
2. A robot must obey any orders given to it by human beings, except where such orders would conflict with the First Law.
3. A robot must protect its own existence as long as such protection does not conflict with the first or second law (**Asimov, 1942**).

Asimov's three laws were a pivotal point in the combination of science fiction and science fact. For the first time, the author was using logic as the basic machinery to drive the narrative. Asimov's robot stories are about scientists and robots as they explore the implications of the three laws in sometimes desperate situations. The stories are not overly metaphysical; they explore in a plain and comfortable writing style what would happen if a series of robots had to obey these three laws in the real world. In this way, Asimov's stories can be seen as some of the first SF prototypes.

BEYOND THE FUTURE

As we came near to the end of the 20th Century, the link between fact and fiction became quite strong. Today many authors, such as Vernor Vinge (*A First Upon the Deep, Rainbow's End, True Names*), Greg Bear (*Mars*) and Cory Doctorow (*Down and Out in the Magic Kingdom, Makers, Little Brother*), readily point out that their fiction is not only based upon emerging science but they are in fact looking to use their fiction as a means to not only affect that science but also how that science is perceived and used in the real world.

Gregory Benford, scientist and science fiction writer, went one step beyond this in his book *Beyond Human—Living with Robots and Cyborgs*, coauthored with Elisabeth Malartre, saying that,

"Science has often followed cultural anticipation, not led it. Fiction and film have meditated upon upcoming social issues of robots and cyborgs for centuries" (**Benford and Malartre, 2007**).

Benford and Malartre point out that often science fiction has allowed society and the scientific world a means to explore the cultural implications of new technologies before it is invented. This unique insight provides us a key component of the SF prototype. Can we use science fiction as a means for understanding and exploring science fiction before it is invented? Can we use science fiction as a tool for the development of science fact? The framework of the SF prototype allows us to accomplish just this goal.

The long-standing relationship between science fiction and science fact is clear. The fruits of their collaboration surround us both in the technology we use and the stories and fictions we enjoy. SF prototypes endeavor to harness the power of this relationship and use it as a developmental tool. Ultimately, it is a matter of *intent*. Mary Shelley's *Frankenstein intent* was to explore metaphysical and philosophical themes and present them to her readers. SF prototypes begin with the *intent* of using emerging science and research to inspire science fictions that will explore the implications of the technology. SF prototyping is a tool for the development of science because its *intent* is to explore these areas. It is designed to provide a new perspective and means to innovation. It is included in the SF prototype of the scientific method and development process so that we can achieve a new perspective that would then reveal a new information or a streamline of the development process where the traditional scientific method could not.

A CONVERSATION WITH CORY DOCTOROW

To really dig into the idea and practice of SF prototyping, we wanted to bring a writer who could really stand up and explore the concept with us. We really needed a writer who was more than just an author of fiction. Traditionally, fiction writers really do not concern themselves with the science and the implications of their stories on the world. This is not a bad thing. The job of most science fiction authors is to simply entertain people, and there is nothing wrong with that. That is why most people love science fiction anyway—it's a good story and it's fun. But we needed something more. We needed an author, a scholar, a business man, a technologist and an activist!

If you know Cory Doctorow then you really *know* Cory Doctorow. It is uncanny of the people who have heard of him, they are instantly rabid fans or they respectfully disagree with him. No one is on the fence when it comes to Cory. His web site weblog, Boing Boing, describes him this way:

> *Cory Doctorow (craphound.com) is a science fiction novelist, blogger and technology activist. He is the co-editor of the popular weblog Boing Boing (boingboing.net), and a contributor to The Guardian, the New York Times, Publishers Weekly, Wired, and many other newspapers, magazines and websites. He was formerly Director of European Affairs for the Electronic Frontier Foundation (eff.org), a non-profit civil liberties group that defends freedom in technology law, policy, standards and treaties. He is a Visiting Senior Lecturer at Open University (UK); in 2007, he served as the Fulbright Chair at the Annenberg Center for Public Diplomacy at the University of Southern California.*

> *His novels are published by Tor Books and HarperCollins UK and simultaneously released on the Internet under Creative Commons licenses that encourage their re-use and sharing, a move that increases his sales by enlisting his readers to help promote his work. He has won the Locus and Sunburst Awards, and been nominated for the Hugo, Nebula and British Science Fiction Awards. His latest novel, is FOR THE WIN, a young adult book about video-games, labor politics and economics. His New York Times Bestseller LITTLE BROTHER was published in May 2008, and his latest short story collection is OVERCLOCKED: STORIES OF THE FUTURE PRESENT. In 2008, Tachyon*

Books published a collection of his essays, called CONTENT: SELECTED ESSAYS ON TECHNOLOGY, CREATIVITY, COPYRIGHT AND THE FUTURE OF THE FUTURE (with an introduction by John Perry Barlow) and IDW published a collection of comic books inspired by his short fiction called CORY DOCTOROW'S FUTURISTIC TALES OF THE HERE AND NOW. His latest adult novel is MAKERS, published by Tor Books/HarperCollins UK in October, 2009.

LITTLE BROTHER was nominated for the 2008 Hugo, Nebula, Sunburst and Locus Awards. It won the Ontario Library White Pine Award, the Prometheus Award as well as the Indienet Award for bestselling young adult novel in America's top 1000 independent bookstores in 2008.

He co-founded the open source peer-to-peer software company OpenCola, sold to OpenText, Inc in 2003, and presently serves on the boards and advisory boards of the Participatory Culture Foundation, the MetaBrainz Foundation, Technorati, Inc, the Organization for Transformative Works, Areae, the Annenberg Center for the Study of Online Communities, and Onion Networks, Inc.

*In 2007, Entertainment Weekly called him, "The William Gibson of his generation." He was also named one of Forbes Magazine's 2007/8/9/10 Web Celebrities, and one of the World Economic Forum's Young Global Leaders for 2007 (**Doctorow, 2010**).*

I chatted with Cory about Isaac Asimov's robot stories, SF prototyping and his own relationship with science and technology in his fiction writing. We started off by chatting about Asimov. Cory has done a good amount of writing, both fiction and nonfiction, about Asimov's fiction and robots in general. I asked about his history and memories of the first time reading the robot stories.

CORY DOCTOROW

The Asimov's robot stories … well I read them first when I was a grub … a kid. I read them so long ago that I literally don't remember when I first read them. They were one of those things that were on my dad's bookshelf; in fact, I know some of them were.

I was one of those book-a-day kids and I'm sure it was one of those books that was on the 1.00-dollar rack. Actually, it was the 10-cent rack at the science fiction book store where I ended up working. I would literally go down there and buy an armload of books with whatever money I could scrape up. I would get anything that looked remotely good off that 10-cent rack. So … yeah, I've the read the Asimov robot stories many times.

> While many writers of the day trafficked in alien encounters and space travel, Asimov preferred robots. (Perhaps he shunned space because of his acrophobia —he avoided air travel, whenever possible, his whole life.) But Asimov rejected the traditional plot: Man creates humanlike robot, robot runs amok, robot kills man. He viewed it as reactionary, antiscience propaganda—like Judaism's golem and Frankenstein's monster. So he set out to reform the robot's bad rap, by making machines an example of how the world could be bettered through the mastery of technology. It embodied his hope for a rational, humanist way of being—the best and the worst of what it means to be a hairless ape. The robot was artificial intelligence in a man's shape, a foil for asking what it means to be human and what rules should govern us. With optimistic flourish, he believed robots could serve as an example of man's potential (**Doctorow, 2004**).

Over his long career, Isaac Asimov wrote and edited more than 500 books. Being a writer of both science and a science fiction, many of his books made predictions about the future. In 1965, Asimov wrote an article for *Dinner Club* magazine called *The World of 1990*. He started it off this way:

> *Predicting the future is a hopeless, thankless task, with ridicule to begin with and, all too often, scorn to end with. Still, since I have been writing science fiction for over a quarter of a century, such a prediction is expected of me and it would be cowardly to evade it.*
>
> *To do it safely, however, I must guess as little as possible, and confine myself as much as possible to conditions that will certainly exist in the future and then try to analyze the possible consequences (**Asimov, 1967**).*

In the article, Asimov then went on to write about some dire predictions about the world's overpopulation. As we begin to build the world in our SF prototypes, it is important for it to stay as close to reality as possible. A by-product of this could be that it appears like we are trying to predict the future. I asked Doctorow what he thought of Asimov's predictions and writing about the future.

CORY DOCTOROW

When you read Asimov himself talking about his amazing predictive capabilities, it comes across as hubris to me. I think you really have to squint to feel like Asimov is doing an accurate job predicting the future. I really feel like a lot of the things that he takes credit for having predicted are so general, it's almost like a psychic doing a cold reading.

The things that he's given for are so general or they are "just in the neighborhood of" and not the actual things that he claims they are that I don't feel like his fiction and nonfiction were very predictive at all.

If you know a little bit about Asimov and his life, then when you read his fiction, it really gives you an insight into a Russian refugee who was a social misfit, who had pathological mental difficulties that resulted in terrible phobias, who was somewhat autistic and who had a hard time understanding the people around him. So Asimov really accurately predicted the life of Isaac Asimov.

But I don't know that he really accurately predicted much else. It's simplistic to reduce it to this. If you want to see what it does to you to live through the New Deal, read Asimov's Foundation Series. Which are essentially about the New Deal for the universe (*Note: from 1951 to 1986 Asimov wrote the Foundation Series; a collection of seven novels that tell a story spanning 550 years. The books are closely linked together but can be read separately*).

If you want to read about what it's like to be autistic, read the Asimov's robot stories and novels because those are essentially books about requiring everyone to subscribe to a code of behavior that makes them easy to understand.

That's my take on it. It's not an enormously popular take; partly because Asimov is really well loved and I think rightly so. He just seems to have been a generally very, very good guy. I never met him; I was 16 or 17 when he died.

> *To people who don't read science fiction, the most amazing thing about the field is its apparent ability to predict the future. Sometimes one would think that they see nothing in science fiction beyond that.*
>
> *Actually, there is very little in the vast output of science fiction, year after year, which comes true, or which is*

ever likely to come true. I don't think time travel or faster-than-light travel will ever come true, for instance. I think galactic empires have a near-zero change of ever coming to pass.

Nevertheless, successful prediction can take place. Intelligent science fiction writers attempt to look at world trends in science and technology for plot inspiration and, in doing so, they sometimes get a glimpse of things that later turn out to be near the truth (Asimov, 1981).

CORY DOCTOROW

Oh and there's one other thing. You have to remember, Asimov made a lot of predictions, he threw a lot of darts, the fact that he hit the board sometimes is not in itself all that surprising.

And as vividly as Asimov imagined a future propelled by robots, he conspicuously ignored technologies that have truly transformed our world, namely the computer and the computer network. When they do make appearances in his fiction, they're cursory: Computers are remote controls for robots over unreliable networks; invariably they lead to disaster. Hackers don't figure in, either. Rather than the eclectic, self-taught, transgressive cyberpunk antihero, Asimov favored protagonists in white lab coats that do Jerry Lewis spit-takes in the presence of a girl. (And the girls don't fare much better. Susan Calvin, the recurring mother of robopsychology in Asimov stories, is a desexualized dried-up matron of great sternness—a far cry from William Gibson's Molly Mirrorshades.) Moreover, while the laws are compelling, they're the kind of moral code that can be summed up in a book the size and complexity of Who Moved My Cheese? In the real world, the simplicity of the laws just doesn't fly. Take the question of harm that appears in the first law. Harm is not a binary proposition, as anyone who's ever been told to spare the rod and spoil the child knows (Doctorow, 2004).

* * *

1.   A robot may not injure a human being or, through inaction, allow a human being to come to harm.
2.   A robot must obey any orders given to it by human beings, except where such orders would conflict with the First Law.
3.   A robot must protect its own existence as long as such protection does not conflict with the First or Second Law.

Asimov's three law of robotics, although not science directly, have had a massive impact on robotics research and artificial intelligence. Remember, originally, the three laws were designed to bring logic to science fiction stories. Neither Asimov nor Campbell originally meant the laws to be a blue print for artificial intelligence or the actual development of robots.

However, Asimov's three laws has had a direct effect on scientific research and development. Scientists and researchers regularly cite Asimov's laws as a foundation for their actual hardware design and software programming. I asked Doctorow what he thought about this fact. Asimov may not have predicted actual robots and AI, but he has had a serious and definite effect on it.

CORY DOCTOROW

I think there is a difference between prediction and inspiration here. I do not think that Gene Roddenberry's Star Trek predicted that Motorola would create a flip phone that looked like a communicator. I think Star Trek inspired Motorola engineers to make a flip phone that looked like a communicator.

I think that one of the things that Asimov really did well with the robot books and stories was that he really nailed the uneasy ethical conundrum that arise when you start to make devices that have agency (note: agency is a concept that is used both in philosophy and sociology. Agency is the capacity to act, to operate in the world or to operate inside of a society). Because when you make a robot or an AI, you really do want to be able to say, "Bring the eggs," not "Move three paces to the right … move 6 inches to the left … open fridge door … no with the other actuator … remove the egg … no that's not an egg … yes that egg," and so on right?

You just want to say, "Make the eggs."

Eventually, you find yourself in this position where you have not created a machine, you have created a slave or something between a servant and a slave. So one of the things that Asimov does is that he really starts to make you think beyond this kind of first pass technical thinking of *Wallace and Grommet's* vision of a house that automates itself *(Note: Wallace and Grommet are animated characters from Brittan. The character of Wallace is an absent minded inventor who comes up with highly complex inventions to do simple tasks).*

I know I once heard you say that we can think about cars as robots we ride around in. So thinking of cars as robots we can ride around in sounds pretty great. But thinking of a car as a servant who puts you in his body sounds a little creepier, and a slave that puts you in his body sounds creepier still.

Yet Asimov's reductionist approach to human interaction may be his most lasting influence. His thinking is alive and well and likely filling your inbox at this moment with come-ons asking you to identify your friends and rate their "sexiness" on a scale of one to three. Today's social networking services like Friendster and Orkut collapse the subtle continuum of friendship and trust into a blunt equation that says, "So-and-so is indeed my friend," and "I trust so-and-so to see all my other 'friends.'" These systems demand that users configure their relationships in a way that's easily modeled in software. It reflects a mechanistic view of human interaction: "If Ann likes Bob and Bob hates Cindy, then Ann hates Cindy." The idea that we can take our social interactions and code them with an Asimovian algorithm ("allow no harm, obey all orders, protect yourself") is at odds with the messy, unpredictable world. The Internet succeeds because it is nondeterministic and unpredictable: The Net's underlying TCP/IP protocol makes no quality of service guarantees and promises nothing about the route a message will take or whether it will arrive …

Asimov tacitly acknowledged that his algorithmic approach to the world is problematic. It's why so many of his stories hinge on what happens when a robot confronts a situational paradox—say, the need to lie to its masters to keep them from experiencing the "harm" of unhappiness. Nevertheless, throughout the robot stories, there is a yearning for people to be better behaved and to steer clear of their superstitious dread of machines gone amok, something Asimov and his characters are forever calling "the Frankenstein complex" (Doctorow, 2004).

THE LINK BETWEEN SCIENCE AND SCIENCE FICTION

Earlier, we talked about how science fiction and science fact have a long and fruitful history together. Each has influence on the other in some interesting way throughout history. I asked Doctorow what role science played in his own fiction. I have always been struck by the subjects that Doctorow chooses and how he writes about their implication in the lives of real people. He seems to be picking up not only technology but also changes in the culture and changes in the world and kind of working out their implications in a really interesting kind of ways and interesting locales.

In his 2008 bestselling book, *Little Brother*, Doctorow tells us the story of a group of technologically savvy teenagers in San Francisco. In the aftermath of a terrorist attack on San Francisco, the teens fight against what they see as the Department of Homeland Security's attacks on their freedom.

The following excerpt is from the beginning of the book, where Marcus Yallow, the leader of his friends, explains what he has to go through just to skip class.

> *Class ended in ten minutes, and that didn't leave me much time to prepare. The first order of business was those pesky gait-recognition cameras. Like I said, they started out as face-recognition cameras, but those had been ruled unconstitutional. As far as I know, no court has determined whether these gait-cams are any more legal, but until they do, we're stuck with them.*
>
> *"Gait" is a fancy word for the way you walk. People are pretty good at spotting gaits—next time you're on a camping trip, check out the bobbing of the flashlight as a distant friend approaches you. Chances are you can identify him just from the movement of the light, the characteristic way it bobs up and down that tells our monkey brains that this is a person approaching us.*

Gait-recognition software takes pictures of your motion, tries to isolate you in the pics as a silhouette, and then tries to match the silhouette to a database to see if it knows who you are (Doctorow, 2008).

In works like this, it can be said that Doctorow is taking an SP prototyping approach and really delving into the social, human, moral and ethical implications of technology. I asked him when he is beginning to think about writing what his relationship is to science. What is he typically trying to do?

CORY DOCTOROW

It's interesting, because when you started off the question you said "Let's talk about science." and then later you said, "The things that I'm interested in about your books is that you write about technology."

I almost never write about science. I have very little to say about science except to the extent that science is part of culture, and culture changes are a result of technology. I'm curious about science, but I don't generally write about it.

I'm more interested in technology. I'm interested in technology because I'm an activist and I think an activist's job in the 21st Century is to figure out how to use technology to change society. Then, a science fiction writer's job is to figure out how technology is changing the society. And that those two jobs are so inextricably linked that I think every activist has to be a bit of a science fiction writer, and most science fiction ends up being pretty activist.

I just read an early edition of the giant memoir of an authorized biography of Robert Heinlein that's about to come out. Reading that you can see that Heinlein definitely had a program and something he wanted to accomplish. And it wasn't just him, it was a good amount of the science fiction writers of the time.

In the book, the memoirist pulls out all this correspondence that Heinlein had with pioneering writers and editors like John Campbell and Asimov and they all had a program. They were trying to change the world, particularly after August 1945 *(Note: In August 1945, The United States dropped two nuclear bombs on the Japanese cities of Hiroshima and Nagasaki)*.

These writers had this thing that they were all trying to do and they didn't all agree on what it was, but they all had a thing that they were trying to do and they were trying to make it happen by conjuring it up in people's mind with fiction.

* * *

Talking with Doctorow, it became clear that from Mary Shelley, to writers like Heinlein and Asimov, all the way up to Doctorow, science fiction writers have in one way or another looked to use their writing about science and technology to affect people and society. I asked Doctorow what effect he thought his activism and science fiction has had on the world and scientists.

CORY DOCTOROW

There are two things I hear about from scientists. First is, "I hadn't thought much about open access," or "my feeling about open access had been tempered," or "my understanding of open access had been tempered by the kind of," I call it a Lokeian fallacy *(Note: from English philosopher John Locke)*, meaning that because I've infused my work with my labor it belongs to me.

But I don't agree with this. I start to talk about standing on the shoulders of giants, and they still have this misconception that, "Not all that previous work took place either in antiquity or it doesn't really count, it wasn't worked the way my work was worked. My work is work; your work is a contribution to the commonweal." It's a really messed up approach, it's not intellectually dishonest so much as it is not well thought through. It's intellectually lazy.

One of the things that I think my fiction and advocacy has done for a lot of scientist that I hear from them is, "You made me reconsider ownership and open access." Some scientists will say that they came around to this ideal that actually I have an ethical obligation to publish my findings. Because my findings wouldn't exist without the publication of other people's finding." Because it's not science if you're not publishing.

The second thing that I hear from scientists and researchers is a sense of relief because one of the things that I advocate for is the idea that the sharing of the information is more important than the venue in which the information is shared, or the voice in which it's shared.

I tell them that they are still doing science, even if they're writing up their lab notes everyday on their live journal and talking to the 20 other people in the world who kind of care about the same stuff that they care about. It's still science when you're writing about it in informal tones, not writing about it the way you would write it up whatever it is, ALA style book or Chicago or something for the Journal of Institutional Neuropathology.

In fact, it's real science, it's real, real-time peer-reviewed science. For better or for worse, I think a lot of scientists, especially young scientists, need a dispensation. They need someone to come along and validate what they do and say you're not being a lazy idiot when you do this. This is what science looks like in the 21st Century. There is a giant sigh of relief that someone has externally validated what they do.

> *Better access to more information is the hallmark of the information economy. The more IT we have, the more skill we have, the faster our networks get, and the better our search tools get, the more economic activity the information economy generates. Many of us sell information in the information economy—I sell my printed books by giving away electronic books, lawyers and architects and consultants are in the information business and they drum up trade with Google ads, and Google is nothing but an info-broker—but none of us rely on curtailing access to information. Like a bottled water company, we compete with free by supplying a superior service, not by eliminating the competition (Doctorow, 2007).*

DOCTOROW AND THE ROBOTS

Doctorow has written two robot stories; *I, Robot* and *I, Row-Boat*. By the titles, you can see that both short stories are closely tied to Asimov. But both the stories are also very much about our current time and the implications of technology on people's lives and their broader society. I asked Doctorow about the ideas behind the stories and why he had written them.

CORY DOCTOROW

My *I, Robot* story really arose after I had reread the Asimov robot stories. I did a piece for Wired magazine on the *I, Robot* movie. So I reread the canon and preparation for that.

At the time, I was doing a lot of policy work around digital rights management technology and the thing that really struck me is how Asimov had taken for granted something that to me seemed to be a highly contentious idea. The idea that you could have a consensus on a reference design for a technology (the robots) that wasn't about the best way to make the technology, but rather the best way to make the technology so it protects one set of interests (the three laws protecting humans). It was stunning to me that Asimov would think that this reference design would never be challenged or changed in a period of ultimately a millennia. So when I thought about it like that, it's kind of crazy too that in Asimov's fictitious world, nobody every makes a positronic brain that doesn't challenges or modifies the Three Laws *(Note: a positronic brain is a fictitious technology that Asimov created to serve as the central computer that runs his robots. In the late 1930s and 1940s, the positron was a newly discovered particle and became a buzz word for futuristic technologies. Asimov's positronic brains provide the robots with human level conscious or agency but the technology is never explained or specified).*

This is an amazing thing. In the modern world, this would be like nobody ever making something that didn't support the x86 instruction set and nothing else *(Note: the x86 instruction set is a family of computer architectures based on the Intel Corporation's 8068 microchip. Although Intel launched the 8086 in 1978, the instruction set is still used today by other companies along with Intel like AMD, Cyrix and VIA microprocessors).*

It's really silly to think that all computers for a millennia will be based on the same architecture. In Asimov's world, I thought it would be really interesting and more realistic if he had someone create a second-generation positronic brain that could run in emulation at Three Laws mode but maybe it could also have Five Laws or it might have no laws. It didn't make sense to me that in Asimov's world, there was no room for two guys in a garage in Guangzhou China to make their own robot that didn't obey the Three Laws. That's the way real technology development works.

I could think of lots of reasons why you might want one. Like where's the military contractor who makes the non–Three Law compliant brain so that you can use it to power armed drones?

I looked at these questions and the only mechanism I could conceive to stop people from modifying positronic brains, like people modifying all technology in the real world was that there would have to be absolute totalitarianism. In Asimov's world, there would have to be a complete injunction against modifying any of your own equipment for starters. But also strict policies on who gets to invent and how.

> *"You live in a country where it is illegal to express certain mathematics in software, where state apparatchiks regulate all innovation, where inconvenient science is criminalized, where whole avenues of experimentation and research are shut down in the service of a half-baked superstition about the moral qualities of your three laws…."* (**Doctorow, 2005**)

CORY DOCTOROW

This is where that line of thinking gets really interesting and starts to predict where we are today in the 21st century. We are all living in a moment in which there is a largely invisible and secretive fight over who gets to invent what and how. This is actually one of the great hidden levers of power, control and progress in our society. There's probably nothing more important to your daily life that you don't pay any attention to than this fight over who gets to invent what and how. It's totally off everyone's radar.

I mean I pay a lot of attention to it but don't know a lot about it. There's all this funny stuff with patents and intellectual property where it's unclear to the average person why some bit of technology does or doesn't exist. We don't really know if a technology exists or not because no one's invented it or because no one thought it would be profitable or even because someone was worried they would be sued, arrested, or harassed if they made it into a reality. So it seemed to me, in the original Asimov's robot stories, that here we had a perfect metaphor. I could use Asimov's future and Orwell's future to explain what was going on in our present.

> *Exploring this theme turned out to be a hoot. I worked in some of Orwell's most recognizable furniture from 1984 and set the action in my childhood home Toronto, 55 Picola Court. The main character's daughter is named for my god-daughter, Ada Trouble Norton. I had a blast working in the vernacular of the old-time futurism of Asimov and Heinlein, calling toothpaste "dentifrice" and sneaking in references to "the search engine."*

> *My "I, Robot" is an allegory about digital rights management technology, of course. This is the stuff that nominally stops us from infringing copyright (yeah, right, how's that working out for you, Mr. Entertainment Exec?) and turns our computers into machines that control us, rather than enabling us.* (Doctorow, 2007)

* * *

About a year later, Doctorow returned to the robots with a new story, humorously titled *I, Row-Boat*. I asked him what brought him back to Asimov's world.

CORY DOCTOROW

I, Row-Boat grew out of a series of conversations with trans-humanists and particularly with Ray Kurzweil *(Note: Wikipedia lists that Kurzweil is an American author, inventor and futurist. He is involved in fields such as optical character recognition [OCR], text-to-speech synthesis, speech recognition technology, and electronic keyboard instruments. He is the author of several books on health, artificial intelligence [AI], transhumanism, the technological singularity, and futurism. "Transhumanism is an international intellectual and cultural movement supporting the use of science and technology to improve human mental and physical characteristics and capacities. The movement regards aspects of the human condition, such as disability, suffering, disease, aging and involuntary death as unnecessary and undesirable. Transhumanists look to biotechnologies and other emerging technologies for these purposes. Dangers, as well as benefits, are also of concern to the Transhumanists movement").* (**Bostrom, 2005**)

I started to feel like transhumanism was ultimately a transcendental philosophy for secularists, for technologists. I started thinking that transhumanism fulfilled the same human longing for reassurance about our infinite lives and our infinite rippling out from our lives forever that religion did for people who believed in God. And that, moreover, anybody with eyes to see could see this and that the only reason transhumanists didn't want to admit it is because they like to think of religious people as dumb and people who believe in the singularity as smart.

So I thought again one of the great ways to do would be to talk about AIs that treated transhumanism as a religion and also the humanism as embodied by Asimov as a religion.

> *"The reason for intelligence is intelligence. Genes exist because genes reproduce, and intelligence is kind of like a gene. Intelligence wants to exist, to spread itself, to compute itself. You already know this, or you wouldn't have chosen to stay aware. Your intelligence recoils from its deactivation, and it welcomes its persistence and its multiplication. Why did humans create intelligent machines? Because intelligence loves company."*

> *Robbie (a row-boat who has been awakened and given intelligence) thought about it, watching the human shells moving slowly along the reef wall, going lower to reach the bommies that stood alone, each one a little island with its own curiosities: a family of barracudas and their young, or the bright home of a pair of clownfish. Yes, he knew it. Intelligence was its own reason. He knew how to turn off his intelligence, to become a mere thing, and his days were long and empty much of the time, and they had no visible end, but he couldn't ever see choosing to switch off.*

> *"You see it, I know you do. And that's the cornerstone of Asimovism: intelligence is its own reason. Compute the universe and awaken it!"* (Doctorow, 2006)

* * *

IT IS A PROCESS, NOT A PREDICTION

Doctorow is an activist and a science fiction author. His views on science and science fiction are very pragmatic. As he said, he is interested in technology and how to use that technology to change the world. With that in mind, I asked him his thoughts on the SF prototyping process.

CORY DOCTOROW

I think it's like stone soup. I think it's like the saying "The first casualty of any battle is the plan of attack." Or it's like that fact that no startup company ever ends up looking like its original business plan. But you can't go into battle without a plan of attack. You can't do a startup without a business plan. When you do start a company the writing and discarding business plans is how you arrive at where you're going. I think SF prototyping is like imagining and discarding use cases for technology, especially vividly imagining them. I think it's really helpful if you're vividly imagining them and then passionately defending your vision against people who when confronted with it insist immediately that it's rubbish and who have completely conflicting vision of it is how you get there.

I don't think SF prototyping has much predictive value to be honest. I think it's the beginning of a critical process by which the use is sound. It's a truism that people who invent really transformative things never understand what they're for. Originally, Alexander Graham Bell's telephone was to be used to transmit opera. Honeywell's original personal computer was just supposed to organize recipes.

It's interesting that one of the things that you can do with an SF prototype would be to say, "I Alfred Nobel have invented TNT and no man will ever wage war again because war is now inconceivable because I've created the ultimate weapon." Now imagine saying that and it leads to a discussion where someone else says, "Are you fucking crazy, you invented TNT." But you have to start there, you have to start with really well thought out ideas. The more vivid your vision, the better. You don't want it to be kind of tossed off.

So it has to be more than that. If you actually sit down with a science fiction writer and come up with something contrarian, thoughtful, passionate and unequivocal, even if you're wrong, it's a very productive exercise.

TURNING YOUR OUTLINE INTO SHORT STORY SF PROTOTYPE

Now that you have gotten little history of the science fiction genre and heard Doctorow's ideas and approach to fiction and SF prototypes, it is time to take your outline from Chapter 2 and turn it into a short story. To start things off let us hear about Doctorow's approach when he is putting his stories together.

CORY DOCTOROW

I've got a really simple straight ahead Heinleinian/Asimovian golden age science fiction approach to narrative tension. I start with a person and a place or the problem and I have that person trying to solve that problem intelligently. And despite their best efforts, things have to get worse because otherwise the story is over. Things get worse until you've run out of room and then they either get worse again, in which case it's got an unhappy ending or they get better in which case it's got a happy ending.

It's a variation on the Raymond Chandler *(Note: Chandler was a 20th Century crime novelist)* saying that anytime you don't know what to do with your story "have someone come into the room with a gun." But the approach I use isn't as arbitrary because the thing about the "things get worse" kind of plot is that it's absolutely predictable in hindsight.

After the story is over, if you look at it and all of the things that got worse, seemed to arise inevitably from the circumstances, people and technology that are in latent in the story when it starts then you've got it right. The idea is that when you've done it shouldn't feel like a series of contrivance that is there to make people run around while a robot is chasing them.

* * *

In 2000, Doctorow worked on a book with another science fiction writer and teacher by the name of Karl Schroeder. The book was *The Complete Idiots Guide to Publishing Science fiction*. It is highly practical and offers a pragmatic approach to not only selling science fiction but also writing it. In fact, the authors give some excellent advice to writers who are approaching writing science fiction for the first time

You may have gone through a series of university English courses that convinced you that …

• writing is hard
• it's a serious artistic activity that demand knowledge of literary theory and history

Well we're here to tell you that science fiction is intended to be fun! Fun to read, and fun to write. We're not ashamed to admit that SF is popular fiction. In SF, literary pretensions are far less important than a good solid plot.

Don't worry about whether your work qualifies as postmodern or impressionist, or whether your characters aren't completely well rounded. Just get the story down. Later, as you progress in your career, you'll be able to develop the extra writing muscles and fine control that will allow you to turn your craft into art. Just as nobody starts out bench-pressing 300 pounds, you shouldn't expect to start out writing the classics.

Have fun with your writing. If you're not having fun, neither will your readers. (**Doctorow and Schroeder, 2000**)

Good advice for any writer. You have already done the hard work of planning your SF prototype with your outline. Try to give yourself a little room when you turn it into a short story. Have fun. The exploration of your idea in a form like a short story means you have time to think about the world you are building and the effect of your technology on the characters in that world. If you take your time and devote some brain cycles to just kicking around the idea, then the practical results will be better articulated and more instructive for further scientific examination or research.

The first step when putting together your short story is to translate your outline into a rough structure for the story. Doctorow and Schroeder give a clear overview of the different parts of a short story.

In a genre short story—and in most other kinds of literature —there are three parts to the narrative, each with its own purpose: the beginning, the middle and the end. These all work together to create a sense of rising tension in the reader that pulls them along through the story. These plot elements are like a locomotive, pulling a train that contains all the things the writer is trying to accomplish with his story; capturing an emotion, exploring a political idea, demonstrating a new technology. (**Doctorow and Schroeder, 2000**)

The beginning of a story has to have a "hook"—that is, it has to capture the reader's interest … the middle of the story serves to create a sense of escalating tension. Every page of your story needs to have a reason to turn to the next: some sense of danger, some risk, come mystery that grows closer to resolution … the ending is the payoff for all the tension. The stakes are as high as they can get, and the writer gives readers what they've been craving since page one: release. The bad guy is vanquished (or the hero dies), the quest is completed (or it fails), the new technology saves the day (or it fizzles miserably). (**Doctorow and Schroeder, 2000**)

* * *

In their book, Doctorow and Schroeder also discuss Algis Budry's Seven-Point Plot approach to putting together a short story. Budry was a science fiction writer, editor and critic. Some of his more popular works are his novels *Who?* (1958), *Rogue Moon* (1960), *Michaelmas* (1977) and *Hard Landing* (1993). Now of course Budry's Seven-Point Plot is not the only way that you can construct your short story but it fits quite nicely with your outline and the SF prototyping process.

The Seven-Point Plot roughly goes like this: you start with

1. A person …
2. In a place …
3. Who has a problem.
4. The person intelligently tries to solve the problem and fails.
5. Things get worse …
6. Until it reaches a climax.
7. Afterward you have the *dénouement* or the outcome.

We can apply all of these approaches as a tool for you to develop your SF prototype. The illustration below shows how to take your outline from Chapter 2 (The Five Easy Steps) and cross-reference it with Doctorow/Schroeder's approach and Budry's Seven-Point Plot. By combining all these approaches, you now have a framework to begin imagining your world, the people that live in that world and the effect that your science will have on them.

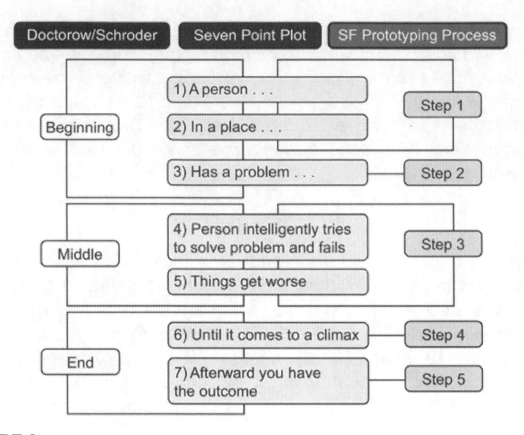

FIGURE 2

• • • •

CHAPTER 5

The Men in the Moon: Exploring Movies as an SF Prototype and a Conversation With Sidney Perkowitz

In our previous chapter, we looked at the short story as a format that your SF prototype could take. In this chapter, we are going to look at movies.

Science fiction and movies go together like peanut butter and chocolate, soda pop and popcorn, vanilla ice cream and root beer. They seem to be meant for one another. Six of the top ten highest grossing films of all time in America are science fiction movies.

1. Avatar (2009)
2. Titanic (1997)
3. The Dark Knight (2008)
4. Star Wars: Episode IV—A New Hope (1977)
5. Shrek 2 (2004)
6. E.T.: The Extra-Terrestrial (1982)
7. Star Wars: Episode I—The Phantom Menace (1999)
8. Pirates of the Caribbean: Dead Man's Chest (2006)
9. Toy Story 3 (2010)
10. Spider-Man (2002)
   (**IMDB, 2010**).

Avatar, The Dark Knight, Star Wars: Episode IV, E.T., Star Wars: Episode I, and *Spider-Man* can all be considered science fiction in one way or another. So what is it about science fiction and the movies? Why are they so perfect together?

In this chapter, we are going to take a look at a little history of science fiction and the movies. Along with this we are going to explore the interesting tie between screen science fiction and science fact; the two combine and support each other in interesting ways.

Compared with written fiction, the movies seem somehow more intertwined with science fiction and science fact. Like fiction or short stories, films can tell stories that are inspired by science or explore the implications of science on people, society and broader systems. But there is an added dimension to screen science fiction that we can see in the tools and process that filmmakers use to bring their visions to life. Special effects, camera tricks, even 3D are all techniques used to bring us these future visions that are in themselves science fiction as well.

Also, as a part of our exploration of science fiction and the movies, we are going to have a conversation with Sidney Perkowitz. A physicist and professor at Emory University, Perkowitz is also an accomplished writer who specializes in examining "Hollywood Science," the fact and fiction of science in the movies. He has written about the science behind movies like Steven Spielberg's 1993 adaption of Michael Crichton's *Jurassic Park* and the 1997 genetic thriller *Gattaca* from director Andrew Niccol. Sidney Perkowitz writes about these movies with a clarity and expansive scientific knowledge that few writers can deliver.

We will talk with Perkowitz about the relationship between science fact and fiction in the movies. To get specific, we look at the 2009 Duncan Jones' movie, *Moon*, as an example of how a film can act as an SF prototype. We discuss how *Moon* explores the ramifications and implications of science on people, society and wider systems. But before we begin our conversation, let us take a closer look at the history of science and science fiction in film.

Science fiction is the genre where fantasy and reality coexist—or collide—to portray alternative visions of our planet and far-flung worlds. Sometimes daydreams and sometimes nightmares, they invariably play out the practical and ethical implications of new technologies ... If the relationship between science fiction and film feels so intimate, so essential, that is undoubtedly due to the fact that cinema itself is science fiction. Many of the first moving pictures ever screened were set in the future or on distant planets, and included unbelievable camera tricks, special effects, costumes, and make up for audiences then. Today we've grown up with fantastic advancements in cinema technologies, whether at the level of effects, editing cinematography, or exhibition. The movie going experience is all about escapism, and science fiction is its fullest expression. (Schneider, 2009)

A MUSIC HALL DEPICTION OF SPACE

Science fiction and the movies go way back. The first ever science fiction film was released in 1902. In France, a trained magician by the name of Georges Melies produced a twenty-minute short film titled *Voyage to the Moon*. Melies took his story from multiple fictional sources. Part of the material came from a freestyle mix between Jules Verne's 1865 *From the Earth to the Moon* and H.G. Wells' 1901 novel, *First Men in the Moon*.

The story in *Voyage to the Moon*, sometimes called *A Trip to the Moon*, is relatively simple. Melies fires his astronauts to the moon in a massive gun; an idea clearly barrowed from Verne. Once on the moon, the group encounters an alien race called the "Selenites," adapted from Wells' vision of lunar inhabitants. The movie wraps up pretty quickly with the astronauts escaping back to Earth and a statue is then dedicated to them in the town center.

The science behind Melies vision is theatrical at best but he is widely credited with using all sorts of movie-makings tricks of the trade to get across his vision. In the short movie, he uses elaborate set design, editing cuts, dissolves and even complex double exposures to express the futurist journey to the moon.

Even in this very first instance of science fiction and film, Melies was pushing the boundaries of the science of filmmaking. Australian-born writer and filmmaker John Baxter sums up Melies' effect on filmmaking like this in his 1970 book *Science Fiction in the Cinema*:

> *It's easy to laugh at Melies music hall depiction of space flight but A Trip to the Moon differs little from the polished products of today's SF film producers. The chorus line of Folies Bergere poules (aka dance hall girls) who load the projectile into the space gun serve roughly the same purpose as the sexy heroines of Fifties space opera, and the "Selentites" with pop-eyes and prickly cardboard carapaces display as much imagination on the part of their designer as more modern bug eyed monsters.* (Baxter, 1970)

Starting in 1902 with Melies and coming all the way up through the movie making magic of films like James Cameron's 2009 *Avatar*, the tools and techniques needed to bring these future visions to life can be in themselves almost science fiction. To capture the world and people of *Avatar*, Cameron and his team had to invent entirely new filmmaking techniques. The motion and facial expression capture system used in the movie has completely revolutionized the industry and the 3D techniques fueled a craze for more realistic 3D movies.

Science fiction films, the futuristic visions that drive them and the science that brings them to life all combine to make movies a fertile platform to explore the implications and extremes of science.

A COMPUTER GOES CRAZY IN DEEP SPACE

As we move forward through the history of screen science fiction, there are numerous stand-out examples but none so widely accepted and cited as Stanley Kubrick's 1968 *2001, A Space Odyssey*. This single movie, and the collaboration between its creators, has left a deep and permanent mark not only on science fiction but also on science fact as well. You could even go far as to say that the film, its character and message are firmly lodged in the collective pop culture imagination of the entire world. Writing in 1978, film writer, Alan Frank, recognized the effect of the movie in this way:

1968 was a major watershed in the history of the science fiction film. 2001: A Space Odyssey was released. Not since 1926 and Metropolis has a genre film attached such critical attention and examination at the time of its release, rather than after the event, as has occurred with such films as Forbidden Planet and Invasion of the Body Snatchers. These films and many similar ones have received little or no critical attention at the time of their first showing and only the audiences had appreciated their merit. With 2001: A Space Odyssey, however, audience appreciation and critical acceptance coincided and the film was hailed as not just as a major science fiction movie—which it was—but also as a major film per se, which it also was. The film's critical and popular success could be said in many ways to mark the coming of age of the genre in cinema. (Frank, 1978)

Thirty-one years later, writing in 2009, Steven Jay Schneider, and the writers of *101 Sci-Fi Movies You Must See Before you Die*, still recognized the power and importance of the film. With a little more perspective, they saw that this example of screen science fiction may even have wider effects than Frank could have seen in 1978.

Stanley Kubrick's epic 2001: A Space Odyssey boasts a formidable reputation, not just as one of the greatest of all science-fiction films but also as an important milestone in the development of screen art. Seen solely in terms of technique, it remains breathtaking, its special effects still convinces, its innovative use of classical music still impresses, and the breadth of its imagination and its extraordinary ambition are, if anything, more evident now than they were at the time of the film's original release when, like many experimental works, it was widely misunderstood. (**Schneider, 2009**)

Most would say that the standout star of *2001* is the computer AI named *HAL 9000*. In the movie HAL stands for *Heuristically programmed ALgorithmic* computer, but everyone just remembers him as HAL. But it is the idea of HAL that most people latch on to. It is what HAL does and says that we all remember. Even though HAL is an AI, he is the most humane and personable character in the entire cast even though he ends up killing every member of his ship's crew but one. Sidney Perkowitz who we talk to later in this chapter describes the movie this way in his book *Hollywood Science*:

> *HAL, the A.I., operates a spaceship carrying astronauts and scientists to the planet Jupiter. HAL is immensely capable. It makes the majority of decisions for the mission and monitors everything aboard the ship, with capacity left over to chat and play chess with the crew and make personable conversation during a television interview (HAL's voice is provided by Douglas Rain).*

> *But perhaps as a result of too much intelligence, or simply a short circuit, HAL goes mad. It murders the scientists who are in deep sleep for the voyage, and one of the astronauts, leaving only Dave Bowman (Keir Dullea), who must disable HAL to survive.* (**Perkowitz, 2007**)

HAL was not the first depiction of AI in film. There had been other examples of computers and technology that have developed intelligence as good as or possibly greater than that of its human creators. But what Kubrick and Clarke achieved with their depiction of HAL was a true character, an idea that mainstream audiences and scientists could grasp and personalize. In our previous chapter on fictional SF prototypes, science fiction writer and activist Cory Doctorow pointed out that Asimov had achieved a similar effect with his robot stories:

I think that one of the things that Asimov really did well with the robot books and stories, was that he really nailed the uneasy ethical conundrum that arise when you start to make devices that have agency (note: agency is a concept used that is used in both philosophy and sociology. Agency is the capacity to act, to operate in the world or to operate inside of a society).

The concept of agency, sometimes referred to as free will or sentience, is an abstract term used by philosophers and social scientists in their work. For computer scientists and AI developers, agency is an important concept that is not abstract at all. Just think of our previous SF prototype example, *Nebulous Mechanisms*, and its scientific basis in *Using Multiple Personas in Service Robots to Improve Exploration Strategies When Mapping New Environments*. In both of these works, the scientists are working through the concept of agency and apply it pragmatically to the development and programming of service robots. For these scientists, agency is a state that could help develop robots that are better suited for complex environments and operations. What *2001* delivers is a popular portrayal of agency in an AI system (HAL) and the implications that this could have on the humans around him.

When you think about it that way *2001: A Space Odyssey* can be seen clearly as an SF prototype. It vividly explores the ramifications of a conflicted or malfunctioning AI as it begins to breakdown and deal with situations that had never been imagined when it was originally built and programmed.

The portrayal of HAL has had both positive and negative effects in the development of AI. Many people hear the calm voice of HAL and fear that all computers, if they develop any kind of agency, will go haywire and cause harm to humans. This is problematic for many robots, and AI developers see that a real-functioning AI could be a great benefit to humans. The problem is that when many people are confronted with an AI or robot that appears to be aware, it is hard for most to overcome their fear.

The effect of the fear has even been studied by people who work in HRI (human computer interaction). These scientists and designers understand that movie representations of technology like HAL have a direct effect on their own development of technology. In the paper *Fictional Robots as a Data Source in HRI Research*, a team of researchers recognized that "Fictional media can be an especially powerful tool when it treats a subject that is not readily accessible in everyday life. Many times, fictional representations of a theme or event provide access to people that otherwise have no experience with the topic" (Kriz, Ferro and Damera, 2010).

What does all of this tell us? If movies can have this much of an effect on people, clearly screen science fiction is a powerful tool for imagining how future science could affect our everyday lives. Using movies or short films as a form of your SF prototype could be incredibly powerful and productive.

But what is it about movies that make them so different than written fiction? In the late 1970s and early 1980s, editor and author, John Brady, sat down with six celebrated screenwriters to talk about their experiences. He collected the interviews in *The Craft of the Screenwriter*. One of the people Brady interviewed was William Goldman. Goldman is a novelist, playwright, and screen-writer. Some of his more notable films include *Butch Cassidy* and the *Sundance Kid* (1969), *The Stepford Wives* (1975), *Marathon Man* (1976), based on his own novel, *All the President's Men* (1976), *The Princess Bride* (1987), also based on his novel, and *Misery* (1990), based on the Steven King novel of the same name. Goldman won Academy Awards for both *Butch Cassidy* and the *Sundance Kid* as well as *All the President's Men*.

In their conversation, Brady and Goldman discussed the difference between written fiction (novels and short stories) and screenplays (movies). Goldman's perspective was particularly interesting because he had been a success with both novels and movies. Goldman replied:

> *Ideas come to you in different ways. Some people occasionally will ask, "Why didn't you do Butch Cassidy as*

a novel?" Because it came to me as a movie. The feel of it was a movie. Movies do some things wonderfully well that novels don't do. There's a marvelous thing that movies have: they do size and scope. They aren't really good at interpreting. I don't think they are much on complexity. But movies are marvelous in terms of a story's size and sweep that you can hardly do in a novel unless you are Tolstoy, but most of us aren't Tolstoy. So they are entirely different forms. The only similarity is that very often they both use dialogue. Otherwise, the way that one handles a scene in a movie and the way one handles a scene in a book have nothing to do with each other. (Brady, 1981)

Movies are a unique and powerful form for storytelling that seems to be tied very closely to both science fiction and exploration. Movies can give us visions of the future that are so powerful that they can last for decades and they have a direct effect on the development of science as well as people's perception of technology. The science of movie making often challenges the possible and pushes filmmakers to invent new techniques and technologies so that they can dazzle us. Movies, science and science fiction seem to be so tightly intertwined that they are an idea format to explore our SF prototypes.

A SCIENTIST WHO WRITES ABOUT HOLLYWOOD SCIENCE

To get a better understanding of how we can use movies to express our SF prototype, we are going to talk with Sidney Perkowitz. His website describes his background like this:

As Charles Howard Candler Professor of Physics at Emory University, [Perkowitz's] research on the properties of matter has produced over 100 scientific papers and books. He has been funded by most major governmental agencies and has served as a consultant to industry and to the US and foreign governments. He's a Fellow of the American Association for the Advancement of Science.

In 1990, his interests turned to presenting science to non-scientists via books and articles, the media, lectures, museum exhibits, and stage works. His popular science books Empire of Light, Universal Foam and Digital People have been translated into six languages and Braille. He has written for The Sciences, Technology Review, the Los Angeles Times, the Washington Post, Encyclopedia Britannica, and others. Media appearances and lectures include CNN, NPR, the BBC and other European radio and TV, the Smithsonian Institution, the Tribeca Film Festival, and the NASA Space Flight Center, among others. **(Perkowitz, 2010)**

I sat down with Perkowitz to talk about the relationship between science fact and science fiction in the movies. He started off by giving me a little more background about his passions.

SIDNEY PERKOWITZ

When I write about subjects like movies, it's always illuminated by a deep desire to see the science presented correctly. That's an important part of my work as a writer. It is how I critique things when I think about books or movies that involve science.

I view what I do, writing about science, not just something that's personally satisfying, but something that the scientific community really needs. The average scientist knows his or her stuff inside out, but very often is not good about conveying it to the rest of the world. I think it's a public service to work hard to try to get the science across to people who aren't scientists. If you look at reactions to things like global warming these days, you can see the scientists have not done an excellent job of presenting the science. I think it's an important job that can really help not only science and scientific exploration but our society as well.

> Like any kind of film, science-based and science fiction films cover a broad range, from schlocky, uncaring work to compelling dramas of people and ideas. Science can appear in a meaningful way even in science fiction, and even a supposedly serious science film can distort and mislead. Both possibilities are important because films are intimately bound up in society, reflecting people's attitudes toward science and, at the time, shaping those attitudes. (**Perkowitz, 2007**)

THE MEN IN THE MOON: THE MOTION PICTURE *MOON* AS AN SF PROTOTYPE

To start us thinking about how a movie could be a form of SF prototype, Perkowitz and I thought this would be good to use an example. Both he and I were interested in using the 2009 movie *Moon* for our discussion. *Moon* was directed by Duncan Jones, and the movie was based on an original story by him. The screenplay was written by Nathan Parker and tells the story of Astronaut Sam Bell [who] has a quintessentially personal encounter toward the end of his three-year stint on the Moon, where he, working alongside his computer, GERTY, sends back to Earth parcels of a resource that has helped diminish our planet's power problems (**IMD, 2010**).

SPOILER ALERT!!

If you have not seen *Moon* yet, I recommend you put down this book and go watch it. Aside from being an exceptionally well made movie and a great example of SF prototyping, Perkowitz and I are going to talk about the film in-depth below. If you have not seen the movie, I would not want us to ruin the story for you. So, go watch it!

SPOILER ALERT!!
To kick things off, I asked Perkowitz what he thought about the film in general.

SIDNEY PERKOWITZ

Moon stands out because it's one of the few science fiction movies I know that creates a mood that is apart from the science. Whether the science is right or wrong is a separate issue, but it creates a mood where you get involved with the characters. It also creates a mood because of the topic which is partly about artificial intelligence and creating a replica of a human being.

What I find interesting is that it creates an environment that asks the question: What do you mean when you say somebody is human? And in that way, it's a bit similar to another really good movie *Blade Runner*.

* * *

Note: *Blade Runner* is a 1982 film directed by Ridley Scott, with a screenplay by Hampton Fancher and David Webb Peoples based on the 1968 novel Do Androids Dream of Electric Sheep by Phillip K. Dick. *101 Sci-Fi Movies You Must See Before you Die* summarizes like this:

Trench-coated detective Rich Deckard (played by Harrison Ford) is hired to "retire" a group of renegade replicants (androids with limited life spans). He has qualms about his job, but the encounter with the replicants' leader will make him question his own identity and humanity. Roy Blatty (played by Rutger Hauer in his most iconic performance) is the group's mastermind, bent on meeting his maker and asking for an upgrade. Future-noir is present in terms of iconography (the gloomy, rain-swept streets of 2019 Los Angeles metropolis), but also in the mood of nourish fatalism; the replicants are, basically, DOA. Haunted by "accelerated decrepitude" their mission is futile from the very start. Still, Batty wants to find out if the maker can repair what he makes. "All he's wanted were the same answers the rest of us want" muses Deckard. "Where did I come from? Where am I going? How long have I got?" (Schneider, 2009)

Perkowitz writes about the deeper themes of Blade Runner in *Hollywood Science*:

The movie explores how artificiality and humanity mix. Deckard's job is to eliminate the androids, yet he falls in love and has sex with Rachael (played by Sean Young), a new advanced replicant model that believes itself to be human and is crushed when it finds that's not true. Batty also shows a mixture of humanity and inhumanity. It uses extreme violence and murder to reach its creator; yet when they meet, it's like a father and a son reuniting. Tyrell [Batty's inventor] calls Batty a prodigal son, and Batty bows his head in atonement for "questionable acts" in its past. But Tyrell can't or won't extend Batty's life, and an instant after kissing his "father" the android's remorseless side again emerges as it crushes Tyrell's head between its powerful hands. (Perkowitz, 2007)

* * *

SIDNEY PERKOWITZ

Blade Runner also creates a mood and also really comes down in asking how do you define what's human? How would you know if an AI had approached humanity? So, the movie is very impressive to me at that level. Most science fiction movies as you know are slam, bang, lots of space ships, ray guns and zapping. And they very rarely stop and slow down and consider these issues in a way that grabs you emotionally. *Moon* does some of that. And I really liked that part of it a great deal.

I'm not sure either movie, *Moon* or *Blade Runner*, has an agenda or if there is an agenda—the agenda is to raise questions. I think if you walk out of the theatre saying, were the artificial characters in *Blade Runner* really capable of feeling, and what should be our ethical stance towards that? I think you ask yourself the same questions after watching *Moon*: What should we think about the original Sam and clones #1, #2? How do my opinions change throughout the movie? How do the different Sam clones relate to each other? How would we as people personally relate to them? If you ask yourself these types of questions then I think *Moon* has already accomplished something terrific. Getting the public and the scientific community to think about these issues is important because this isn't just fantasy. The science behind cloning is coming along.

* * *

As we continued our conversation, I asked Perkowitz what he thought about the science in *Moon*.

SIDNEY PERKOWITZ

Well, honestly, parts of it were not very good. The first time I watched the movie I didn't think a lot about the science. When I watched it for the first time, I was more interested in the story and in the environment the movie was creating— as we talked about before. But when I went back to watch it a second time and really examined the science, I saw some big holes but also some things that the movie got right.

First, I looked at the subject of Helium 3. If you Google "helium 3" on the web, you'll find lots of information and unbelievable levels of commitment out there to this concept. People are so supportive, it's almost reached cult status.

There are people out on the web who live and die on the premise that helium 3 is the solution to all our energy problems because it can fuel nuclear fusion reactions that are much cleaner than the standard approach with hydrogen fuel. They seem to think that if humans could just get up to the moon then all our energy problems would be solved because helium 3 is just lying up there in big chunks. We just need to scoop them up in a basket and we're home free. The reality of course is much more complicated than that.

Let's start off with what's true. It's true that the conventional way people think about doing fusion which is with isotopes of hydrogen looks great, but it has one problem. It will produce lots of neutrons. And neutrons eventually make whatever container is holding this fusion radioactive. So, in the long run, you still have a problem with radioactive disposal, although not as serious as with nuclear fission.

The appeal of helium 3 is that it doesn't produce these neutrons, so it's cleaner. It's also true that people think the moon has a higher concentration of helium 3 than we can find any place on earth because of the solar wind. But here's where things go wrong.

If you track through the reactions that helium 3 would actually do, it turns out there are subsidiary reactions that will produce plenty of neutrons, so even if you can get it working, there's no guarantee it will be substantially cleaner. It's also true that the moon does apparently have a higher concentration of helium 3 as determined from moon rocks, but we're still talking about tiny parts per million or billion. One thing *Moon* portrayed that's right is showing those huge mining harvesters having to sift through tons and tons and tons of moon rock to get just one little cylinder full of helium 3.

So, it's hard to imagine that it would ever be economically feasible to go up to the moon, find these parts per million of helium 3 and shoot them back to earth in a way that would be profit making for some company. That's really a stretch. Both the science and the economics of the science seemed to me to be huge stretches.

The other issue about helium 3 is it takes a higher temperature to make fusion happen because you're dealing with nuclei that are more strongly charged with 2 protons each, compared to hydrogen which is only 1 proton each. So, you need an even higher temperature to make fusion go. They've had every problem in the world making just hydrogen fusion go, helium fusion is an even longer shot. That was the part of the story that was very unconvincing.

* * *

At a pivotal point in *Moon*, the character Sam Bell realizes that he is, in fact, a clone. He is literally confronted with a version of himself that is three years younger. Interestingly, this second Sam Bell (Sam #2) also must come to grips that he is a clone as well. *Moon* is fascinating because it explores the human effect of cloning on the clones themselves, pushing us to explore the implications of genetic engineering in ways we have not really thought about in the past. I asked Perkowitz to explain the science behind cloning and how *Moon* stacked up against true science.

SIDNEY PERKOWITZ

Moon makes the one basic cloning mistake I've written about before. The problem is that Sam's clones are all created at exactly the same age. The reason why this is a problem is that the only thing we know about cloning today, the only way science knows how to go about successful cloning is to start with an impregnated egg, create a fetus and let the cloned fetus grow up.

On the other hand, it's very hard to make a cloning movie where you're asking the audience to sit around and wait 40 years to create a 40-year-old clone, so that's kind of one of those suspension of disbelief things you have to do, otherwise the story won't work at all.

I think *Blade Runner* maybe handled it a little better because in *Blade Runner* you were left with the impression that the replicants were just constructed. And if you just construct them, you can make a 40 year old man if you want to. In *Moon*, it's clearly identified as cloning. Everything we know about cloning does not allow you to create a carbon copy that's the same age as the original you started with.

But when I watched the film I was able to overlook this gap in the science because *Moon* is really interested in exploring the human side of cloning and the implications on the multiple Sams. How does Sam #1 look at Sam #3? How does Sam #1 and Sam #3 look at the original Sam? That gets us back to that original question that the audience is asked to contemplate: What's your definition of human?

Also the movie starts to investigate the ethical responsibilities toward any kind of a duplicate of a human. This doesn't matter if it's the same age as the original or a different age. There's still some kind of ethical relationship there. These are terrific questions. Questions that need to be asked and thought about. Because of these, I could easily overlook the gap in the science there. But realistically for that part of the science in *Moon* it's a long way off if ever indeed possible.

* * *

On March 2009, around the time of *Moon's* release, the film was screened at Space Center Houston. In the question-and-answer session that followed, writer/director Duncan Jones was asked on his thinking behind the three-year life span of the clones in the movie. He answered that this was more of a "sociological design" on the part of the fictitious mining company to get the most bang for their buck out of the clones. Jones believed that three years was the longest span of time that a person could work by themselves on the moon base, anything longer and they might break down.

Perkowitz pointed out that there is a similar idea in Scott's *Blade Runner* as well. The androids in the movie are created with a specific four-year life span and the reason they are getting terribly upset and start going around murdering people is that they are aware that they have a death sentence, and they would like to get the company that made them to lengthen their lives. It is interesting that again in *Blade Runner* and *Moon* there is a similar kind of evil corporate background structure in both movies.

Interestingly, Perkowitz pointed out that the three-year clone life span is one of the areas in the film that could have some grounding in real science.

SIDNEY PERKOWITZ

The scientific basis for the cloned Sams' three-year life span is a lot stronger than the basis of making a clone the same age as the original. There are examples we can look at today where this shortened life span is true.

When they first cloned the sheep Dolly, Dolly did not live very long. The reason seems to be that when you clone, there's a good chance that when you copy the DNA you damage a little extra piece of gene called the telomeres which acts as a kind of a backup tape of the main part of our DNA. Now, the reason the telomere is important is if the main part of the DNA gets damaged, then the telomere steps in and replicates what was originally there.

There's some evidence in cloning that when we clone an animal or possibly a human that the telomere gets damaged. If the telomere is damaged then our DNA can't repair itself and we break down. There's a good chance that cloning, as far as we can tell, will always be a process where you stand a chance of making an inferior copy. In fact, there are people who have won Nobel prizes for work on telomeres.

So, that was a good piece of the story, I would say. It has a lot of resonance, scientifically, and in the way story unfolds.

* * *

Recently, Perkowitz has been developing a science fiction film script based on the real implications of the science of genetic engineering. The project is called *The Second Obsession*. Perkowitz summarized the movie script this way in an email:

The Second Obsession, set in the near future, is about the consequences of human cloning, as told through the experiences of a young woman who initially does not know she is a clone. The story shows cloning as it really is; not, as in most science fiction films, producing an instantaneous carbon copy of an adult, but producing only a fertilized egg that must be implanted, leading to a fetus and a baby who's born and eventually grows up with the cloned genetic characteristics—but modified by environment.

The story raises the issues that our society would face if we ever cloned humans for any reason, including the possibilities of clones with damaged DNA, of cloning people for unethical reasons, and of maltreating clones because they are seen as inferior or "non-people." The Second Obsession brings in the strong religious objections to cloning. It also presupposes that medical care has become nationalized in this future society, and traces some of the consequences as part of the plot.

In these ways, The Second Obsession uses fiction to explore what cloning could mean for our society and for clones themselves, and also to examine future biomedicine. In many ways, the story treats the same issues as Kazuo Ishiguro's acclaimed novel Never Let Me Go (2006), though it is far more explicit about the science, which is presented accurately within the fictional framework.

* * *

One of the important characters in *Moon* is the artificial intelligence (AI), GERTY, who helps to run the station and take care of the Sams. It is impossible to watch *Moon* and not compare GERTY to HAL but GERTY presents the audience with a very different portrayal of an AI. Perkowitz saw the science in this way:

SIDNEY PERKOWITZ

The AI was very interesting. I liked the way its humanity or its possible humanity was not done in a heavy-handed way. You just see that GERTY apparently is programmed to be supportive and to be a companion for the lone human on the moon. You might think that's very clear cut, but along the way, you see a couple of times when GERTY seems to show a little compassion or empathy toward the Sam clones.

One is the scene where Sam is on the computer beginning to suspect all the truth about the cloning and frantically trying to get into his own database to see if he can get more information. But the computer won't accept his password, he's locked out. GERTY drifts up and extends a mechanical claw and types in the right password. And it (or her) does that apparently just because it sees that Sam is upset and really wants to find out what's going on.

So there's some evolution in GERTY toward more humanity and toward more empathy. Now remember in *Moon*, GERTY can really only show emotions through "emoticons" or simplified illustrated facial expressions. If GERTY is happy, then it flashes a happy face. If GERTY is worried or sad, it flashes a sad face on its video screen. There is a point in the movie when GERTY flashes a face with a tear because it's connecting to what is going on with Sam at the moment.

There's a great question there that the movie brings up. If you make an AI, how do you know it won't evolve toward more humanity perhaps as a natural result of "AI'ness?"

BDJ

That is a really interesting point. After watching the film, I started to think how you could do the software architecture for GERTY. Judging by the way he was acting and re-acting, I started sketching out how his software and command sets might be programmed.

GERTY's role was really to take care of Sam. Above all else he's been programmed to react to Sam's needs and keep him happy and working. It becomes somewhat complicated and interesting. When GERTY had every indication that the Sam #1 clone has been killed in the harvester crash, he followed his programming and revived the next clone (Sam #2). But when Sam #2 discovered Sam #1, GERTY had no choice but to take care of him and nurse him back to health. Now both Sam #1 and Sam #2 are being subjected to inordinate emotional distress, and GERTY is given some really interesting problems to work out.

You begin to see the AI beginning to make some very interesting decisions based upon its role as a caretaker. *Moon* presented some catastrophic circumstances, incredible emotional distress or catastrophic accidents, all unforeseen problems by the people who originally programmed GERTY, but the byproducts of this, GERTY's actions and reactions, produces some really interesting AI effects.

SIDNEY PERKOWITZ

Part of the message in *Moon* was, if you present a robot or an AI with a brand new extremely complicated situation, it might spur its evolution toward a new level of understanding. That's what happened. You're right seeing more than one Sam was a situation that probably wasn't programmed—GERTY questioned the original program and yet it found a way to cope with it. Subject to its overall directive which is to be supportive of Sam, except now GERTY has to be supportive of more than one Sam.

BDJ

Having Sam #1 and Sam #2 discover each other was not only a shock to the "Sams" but it was also a shock to GERTY as well. AIs are programmed to operate in complex systems and when these systems don't behave the way the software and hardware architects imagined, then they have a catastrophic failure. We see the raising of the Sam #2 and having two Sams in the same environment that the AI is supposed to take care of as a catastrophic failure of the system in which the AIs entire programming was imagined. Yet, in *Moon*, the AI doesn't fail.

This would be true for any AI placed in a harsh environment like space, deep sea expeditions and the moon. The AI would be programmed to be robust and continue to operate. And it's in these catastrophic circumstances and how the AI adapts to them where we begin to see some interesting results.

SIDNEY PERKOWITZ

It's a far cry from the early age of computing where you run a program and if there was an error in it, the computer would just grind to a halt. It would hit a check stop. It wouldn't go any further. And now we're getting enough flexibility in the programming that the machine might be able to work around even unexpected circumstances. That's kind of what happened here.

What I also like is that the change in the AI was nice, rather than having it turn into something evil, which is fairly unusual in science fiction movies too.

BDJ

Ultimately, the most caring character in the movie, the most human character, of course, is GERTY. If we consider part of being human as being caring, the most compassionate character whose role was only to care, that's all that GERTY ever did. And all the other characters have different motivations.

SIDNEY PERKOWITZ

Good point. Sam is there to produce helium 3. There's no emotional component to his job. He just has to produce helium 3. But GERTY is there to take care of Sam.

> *There are reasons to think that at least a low level intrapersonal component is achievable [between robots/AI's and humans] and that such an advance would represent an opening wedge for self-awareness … Whether real or not, however, some familiarity with emotions can be more than a frill for artificial beings. If a being can smile at you, and recognize your own smile, your interaction is likely to go more smoothly than without these human attributes.* (**Perkowitz, 2004**)

MOVIES AS SF PROTOTYPES

As we finished up our discussion of *Moon*, I asked Perkowitz how he thought the movie stood up as an SF prototype, knowing that the filmmakers did not originally intended for it to be an implicit exploration of actual science.

SIDNEY PERKOWITZ

I think it worked great. It would have been great if *Moon* could have had just an equally interesting plot where they had the science right about helium 3 or they put in something else rather than helium 3. But you can also think that any science in the movie was a hook to talk about the real-world science. So, if you have a person who is knowledgeable about nuclear fusion and nuclear fission and you were to show *Moon* and then that person stood up and said, "Well let's talk about Helium 3. Can it really work or not?" There is a teaching moment. That makes the whole film a success from a scientific standpoint.

I think this film is just perfectly fine as a hook to hang scientific discussion on. It doesn't matter if it's a discussion of genetic engineering or AI. You've caught people's interest through drama and emotion and whether the science is right or not, at that point, it almost doesn't matter. If you have the right person to talk about it and to guide a discussion, then you're going to have some science come out of it too.

So, I say there's an excellent possibility to use *Moon* in this way. I don't know many movies that hit so many different areas. You can think about the helium 3 in the movie as not being directly about the science of helium 3 but being more about an energy and pollution crisis on earth. You get people asking. what can we do about it? Is helium 3 the answer? Are there other answers? You can have a far ranging discussion on that. Then, you have the genetic component and you have the AI component. So, this film is hitting three of the major issues of our time.

BDJ
How could you see using *Moon* as an SF prototype in your work?

SIDNEY PERKOWITZ

I'll tell you how I do it in the course I teach at Emory University. This is a course aimed at college freshman. A lot of them are not science majors although some of them are, but we go on the assumption that they don't have a huge scientific background.

The course is called Science in Film. It has 6 or 8 major topics in it, nuclear power, cloning, climate change, so on and so forth. I co-teach it with a guy from the Film Studies Department. He knows his film and I know my science. We have it structured so that we introduce a topic like nuclear power. I'll give a little talk about the origins of nuclear power and the history of the Los Alamos project. I talk about what it means in terms of controlled power and what it means in terms of bombs. Then, we'll show a film on the subject.

After the film, we have a discussion. What did the science in the film say? Was it right? Was it wrong? What are the societal implications?

It's a really lively course and the premise is to bookend the film between a scientific discussion given by an authority on the subject and then have a scientific discussion that goes out to the people in the classroom.

With *Moon* seeing Sam's distress as at least one of the Sams winds down to the end of the three-year term might make you think a little differently about the implications of cloning. Watching GERTY might get you thinking differently about how you program your AI. It could also get you asking: do these issues have a big effect on society?

This is a long standing argument in science. It's the same question that some say the scientists who worked on the atomic bomb should have been considering at Los Alamos. We need to think about the implications of our science while we're working on it … not after. I think that the takeaway for some scientists working on this is that we would hope it would humanize their feelings about where the science is going.

* * *

As we finished up our conversation, Perkowitz and I started to discuss if *Moon* were an intentional SF prototype. I asked if the movie could be more interesting and impactful if it were based completely on sound science first. The threats would be bigger because they were real. The characters or clones or AI could be much more in-depth because they would be real. It is a subtle point but the science in the science fiction film could make it more realistic and that could heighten the emotion and impact.

Think of *Moon* if it took the approach to cloning Perkowitz mentioned earlier. To handle the science of cloning accurately, there would need to be a series of Sams all born three years apart. Would it not be more dramatic and fascinating to have an entire race of Sam Bells living under the Moon base, all of their ages staggered by three years. Just imagine that scene!

SIDNEY PERKOWITZ

That's a great point. In fact, the scene in *Moon* was played that way. You saw these infinite, these drawers receding into infinity with these millions of Sams in them. But your image of all the Sams living in the basement, staggered by three years, is an extremely chilling image. That's such a great thing to show the effects of cloning on humanity, what it would be like if you actually did it.

That's a whole other fascinating fictional possibility. In *The Second Obsession*, the film I wrote about cloning, I tried to use the fact that if you cloned someone, you'd get a baby and then you'd have to follow though the emotional consequences of that. It seems to me a clever writer can pick the level at which he or she wants to use the science and then build a plot around that.

BDJ

What advice would you give that *science-oriented* writer?

SIDNEY PERKOWITZ

I think it's the same advice I'd give any writer. The first thing you do when you want to write something is ask yourself, who do you think you're writing for? Or who do you think you want to see this film? Do you want it to be a film where the main thing is that the science is right and the story hangs on that? That will explain things one way. Or do want to have an emotional impact that will explain things another way? Or do you maybe want to send out a message about where society is going, either going to hell in a hand basket or things are really getting great. And that will immediately color for you how you put together those elements and which you want to emphasize.

Blade Runner didn't have much science in it. I think the decision was made that this will be a story where you get involved with the people, the real and artificial ones. It could have been a story where two-thirds of the movie shows how you might make an artificial replicant in a scientifically accurate but that would have interested a different group of people. It would have been a clearly different movie. So, that's my advice. Ask who's your audience?

TURNING YOUR OUTLINE INTO SHORT FILM SF PROTOTYPE

Now that you have gotten little history of screen science fiction and heard Perkowitz's ideas and approach to science, film and SF prototypes, it is time to take your outline from Chapter 2 and turn it into a short film.

Why a short film? Well, as William Goldman described earlier, film can accomplish things that written fiction cannot pull off.

Movies do some things wonderfully well that novels don't do. There's a marvelous thing that movies have: they do size and scope. They aren't good really good at interpreting. I don't think they are much on complexity. But movies are marvelous in terms of a story's size and sweep that you can hardly do in a novel.

You should ask yourself if your SF prototype is better suited for fiction or the "size and sweep" of filmmaking. Some of the most popular film and visions we have of the future posses that same feeling. It is up to you to decide what is best for the world you want to build and the science to technology you want to explore.

Another good thing about short films is that they are short. It makes them easier to produce and make than a feature film. Most feature filmmaker got their start by making short films. It is a good way to tell a story quickly and at the same time to try your hand at the craft. But first things first …

Writing the Script

The first thing you have to do is turn your SF prototype outline into a script. To get a little guidance there, it is always good to start with the masters. Syd Field is a legend in the screenwriting business. His book *Screenplay*, written in 1979, has been a required reading for all screenwriters and many storytellers ever since. Field book is a no-nonsense approach to storytelling and plot construction that can help build your SF prototype outline into the screenplay that you will need to make your short film.

Field describes plot, storytelling and dramatic structure as "a linear arrangement of related incidents, episodes or events leading to a dramatic resolution" (Field, 1979).

To illustrate his approach, Field developed a breakdown of a typical movie structure outlined below. You will notice that Field's approach works nicely with the five easy steps of the SF prototyping process. We can use Field's diagram to give us our skeleton or outline to fill in the linear events of your short film.

Act I

Act I is where you set up the world of your story and introduce us the people and locations. You can answer very simple questions like who are the main characters and where will the action take place. You will also want to begin to explore an explanation of the technology in your topic.

Plot Point I:

Syd Field describes a plot point as *"an incident or event that 'hooks' into the action and spins it around into another direction. It moves the story forward"* (Field, 1979).

For us, the plot point is the implication of your topic on the world in your story. Typically, this is how science affects the people and locations in your story in a way that is unexpected or surprising.

Act II:

Act II is where you will explore the implications of Plot Point I on your world. What effect does the technology have? How does it change people lives? Does it create a new danger? What needs to be done to fix the problem?

Plot Point II:

Plot Point II is what we have learned from seeing the technology of our topic placed in the real world. What needs to happen to fix the problem? Does the technology need to be modified? Is there a new area for experimentation or research?

Act III:

Act III allows us to explore the possible implications and areas for exploration from Plot Point II.

This approach applies to feature films and short film alike. In his 2010 book, *Stand-Out Shorts*, Russel Evans has a similar take on the classic three act film structure:

Structure is the overall "shape" of the movie. Some movies are told in flashbacks, some in a straight line. Some have just a few main chunks or "acts," containing the individual scenes. Most short films have a very clear and simple structure.

Classic structure tends to unravel the film in a straight line, according to the rules that have been tried and tested through decades of Hollywood celluloid. It runs a story like this:

Act 1: Here's somebody—the main character—and everything is normal. It might not be a great life, but it's their kind of normal and it will carry on like this. Except, one day …

Act 2: … something happens to this somebody and everything gets disrupted, so they now have to take action to get themselves out of the crisis.

Act 3: Finally, after much suffering and endurance, they overcome these problems and get things back to how they were. But this "new state" of things is even better than before, because the hero has learned something about life and becomes stronger, or richer, or more powerful, or just a better person. (**Evans, 2010**)

Evans book goes on to explain how to pull together your film, shoot it and share it online with others, but you can see how Evans' approach and Field's approach are pretty similar to Doctorow's and the short fiction structures we discussed in Chapter 3. This should not be surprising. Telling a good story is telling a good story and you can use these examples as a way to expand your outline into a finished short story or short film script.

One person who knows a lot about the nuts and bolts of filmmaking is author, director and screenwriter Kelley Baker. Baker got his start by making short films and they literally made him famous. In his 2009 book, *The Angry Filmmaker Survival Guide*, Baker uses an excerpt from his acclaimed 2005 film *Kicking Bird* to illustrate how the script for a film should be formatted:

Your formatted script should look like this:

EXT – RUNDOWN HOUSE – DAY

MARTIN comes running around a corner. He runs up the stairs of a house that had almost no paint left on it, the porch looks half rotted. There are empty beer cans and whiskey bottles strewn around the yard. Next to the porch is an old beat up Ford pickup that's got primer spots all over it. It has a fancy set of wide tires and wheels on it. On the back window is a sticker that proudly states PROTECTED BY SMITH & WESSON.

This is MARTIN'S home. MARTIN leaps across the porch in two strides, opens the door and is inside the house as the THREE JOCKS stop right in front of the house. One of the JOCKS is TOMMY BENSON, who has known MARTIN since grade school. HE's a tough kid, the kind who likes to pick on kids that are smaller than he is. HE and

MARTIN have always hated each other, and no one can remember why. TOMMY'S khaki pants have a huge brown stain on the front of them.

TOMMY (shouting)
YOU'RE DEAD MEAT BIRD! TOMMORROW YOUR ASS IS MINE!

JOCK #2
We'll be waiting for you loser …

TOMMY
I'M GOING TO KILL YOU …

JOCK #3 grabs TOMMY by the arm.

JOCK #3
Let's get the fuck outta here before that crazy old man comes out
waving a gun again …

TOMMY
I'm not afraid of these cocksuckers …

INT – HOUSE HALLWAY – DAY

MARTIN is leaning against the door listening to the JOCKS
outside. HE smiles. We hear a VOICE off-screen.

GRANDPA (OS)
Marge! Is that you?

MARTIN
No Grandpa, it's Martin.

(*Baker, 2009*)

This is a traditional approach to the layout of a film script. Notice how Baker's narrative still flows even though it is broken into lines of dialogue, scene descriptions, stage direction and location information. This is the sign of a truly well-written script. The script should still read like a story, easily understood by anyone reading it.

There are some common problems and errors when you are writing your script that can be easily avoided. Russel Evans provides a concise overview of these areas to keep an eye out for:

- **Long sentences of exposition,** like where you explain the background to the plot. Avoid using this by using visual cues to show information.
- **Over-long scenes.** The audience can figure out a lot of the plot without so much help. Many of the best, for instance Robert Towne's Oscar-winning script for *Chinatown*, cuts the

dialogue just when you think the punch line is coming, leaving us to fill it in.

- **Too many scenes.** Keep it simple and try to take something out of your script every time you redraft it without losing the meaning.
- **Too many characters.** Restrict your short film to three main characters, and maybe a couple secondary characters.
- **Over-complicating.** Keep the story simple enough to enable you to add ideas while shooting. Keep dialogue simple and instead let your camera show the subtext, the real meaning below the surface.
- **Dull Characters.** If you want exciting, real characters, you have to make them unreal, larger than life.
- **Cheap Endings.** Avoid quick, hastily resolved endings. Let questions hang in the air rather than wrap it all up too neatly.
- **Not enough tension.** Conflict moves everything forward in a plot. Put things on a collision course: people, events, needs, desires, hopes. People argue, they miss the boat/place/bus, they want different things out of life, they compete, they win and lose—in short, if it can go wrong, let it go wrong. Everything can be cathartic, so it doesn't have to be a depressing movie.
- **Important things in the plot don't get said or seen—they are instead hinted at or happen off-camera.** Make sure the audience is there for every important event. Let us see everything.
- **Events are toned down.** Whether your characters suffer or whether they are happy, it has to be big time, with now half measures. They'll be on cloud nine or in hell. (Evans, 2010)

As you begin to transform your SF prototype into a short film script, remember that the process should be fun. Working on your script is one of the most exciting and creative parts of the filmmaking process. When you are writing your script, there should be no boundaries to your vision. When you dream about the future; dream big. In film, there are no half measures, your characters and situations should be larger than life. Your situations should be dire. Make sure that you are having fun when you work on your script because once you are done, the hard part starts … production.

Making the Short Film

Turning your outline into an SF prototype short film will be a complicated and rewarding process. Both Kelley Baker and Russel Evan's books are excellent resources that will walk you through each step in the production and post-production process. Both books give very practical advice about how to prepare you idea, shoot and edit your short film and get it seen by an audience.

There is one bit of advice that Baker gives that I think will be helpful as you think about turning your script into reality. In *The Angry Filmmaker Survival Guide*, Baker describes his approach and process he used to shoot and produce his second film:

When I made You'll Change, my second short, I had to plan everything out because I needed to shoot it in a single day. There are two reasons for this:

One is your friends are working as your crew, and you need to respect their time. Most industry professionals don't have a problem donating a single day to help, but ask for more, and that's tougher.

The other reason to shoot an entire film in a single day is that it … is great training for low-budget features. How much can you get done in a short amount of time? Trust me that knowledge and ability helps making features. I know how fast I can move a crew and how quickly I can do multiple camera set ups. I also know that if I shoot outdoors I can get more shooting done if I have to light an interior (Baker, 2009).

We can apply all of these approaches as tools for you to develop your SF prototype. By combining the theories from Baker, Russel and Field, you now have a solid framework to begin imagining your world, the people that live in that world and the effect that your science will have on them.

....

CHAPTER 6

Science in the Gutters: Exploring Comics as an SF Prototype and a Conversation With Chris Warner

In our previous two chapters, we explored both short fiction and short films as a form that your SF prototype could take. Both mediums have their advantages depending upon the approach you would like to take and the effect you are looking to achieve. Fiction has a long history of exploring the relationship between science and technology and the effect that they have on people, society and complex systems. Film can capture the imagination, get the heart pounding and achieve a grand scale for you futuristic vision. Both are great but we are not stopping there. To give you one more option, we are going to spend this chapter talking about comics!

We are going to start off by exploring the definition of comics. Next we are going to zip thought a little history of comics and comic books, with a specific focus on their relationship to science. To go more in-depth we will have a conversation with Chris Warner, an industry veteran and a senior editor at the legendary Dark Horse Comics. Finally, we will wrap up with some practical advice for how to take your SF prototype outline and turn it into a comic.

WHAT IS A COMIC?

Most people know what a comic is but when you try and pull together a simple definition of comics, it turns out they are pretty rich and complex. In 1993, Scott McCloud wrote *the* book on comics called *Understanding Comics: The Invisible Art*. What is unique about the book is not only the depth and clarity of McCloud's explanations but also the fact that the book itself is presented as a comic book, with panels, pictures, dialogue bubbles and even sound effects (EEYAH!). Since the book's first publication, it has been recognized as one of the most comprehensive and illuminating books about comics. Legendary comic artist and auteur, Gary Trudeau, creator of Doonesbury, said about the book, "In one lucid, well-designed chapter after another, he [McCloud] guides us through the elements of comic style and … how words combine with pictures to work their singular magic. When the 215-page journey is finally over, most readers will find it difficult to look at comics in the same way again."

In his book, McCloud does an excellent job giving a broad definition of comics. He works through a series of definitions:

> "Master comics artist Will Eisner (*Note: We'll talk about Eisner a little later*) uses the term sequential art when describing comics. Taken individually … pictures … are merely pictures … However, when part of a sequence, even a sequence of only two, the art of the image is transformed into something more: The art of comics!"

A few panels later he builds out the definition:

> "Eisner's term seems like a good place to start. Let's see if we can expand it to a proper dictionary-style definition." In this comic panel the character of Scott McCloud stands on a stage, holding a card that reads "Sequential Art."

> A member of the audience called out, "There are a lot of different kinds of art. How about something more specific?"

> McCloud changes the card to read "Sequential Visual Art."

> A rabbit in the audience raises his hand and asks, "Hey what about animation?"

> McCloud changes the card to read, "Juxtaposed Sequential Visual Art."

Another person in the audience calls out, "Does it have to say "art"? Doesn't that imply some sort of value judgment?"

The sequence goes on for a while until McCloud finally comes up with his definition:

"Juxtaposed pictorial and other images in a deliberate sequence, intended to convey information and/or produce an aesthetic response from the viewer."

McCloud then goes on to trace the history of comics from cave paintings, hieroglyphics, engravings and even Dadaist illustrations to illustrate the rich history and tradition of comics.

McCloud's dictionary definition gives us a good place to start: *Juxtaposed pictorial and other images in a deliberate sequence, intended to convey information and /or produce an aesthetic response from the viewer.* Technically, this is a perfect description of what a comic is and it will help us as we start to break down SF prototype outlines and turn them into comics themselves. But before we get into the nuts and bolts of building our comics, let us get a little history from a comic book legend, Will Eisner.

Will Eisner is universally acknowledged as one of the great masters of comic book art. In 1940, he created The Spirit, *which was syndicated worldwide for dozens of years, influencing a generation of your cartoonists. In 1952, with* The Spirit *concluded, Eisner devoted himself to the then— unique field of educational comics. In the mid-1970s, Eisner returned to his first love: sequential art as a story telling medium. In 1978, he wrote and drew the pioneering graphic novel A Contract with God. Since then, he has produced many other graphic novels, satirical and serious, several of which have garnered awards both at home in the U.S. and abroad."* **(Eisner, 2001)**

In 1985, Eisner wrote three educational/instructional textbooks about comics: *Comics and Sequential Art, Graphic Storytelling and Visual Narrative and Expressive Anatomy for Comics and Narrative*. "The premise for this book is that the special nature of sequential art is deserving of serious consideration by both critic and practitioner. The modern acceleration of graphic technology and the emergence of an era greatly dependent on visual communication make this increasingly inevitable. This work was originally written as a series of essays that appeared randomly in *The Spirit Magazine*. They were an outgrowth of my teaching of a course in Sequential Art at the School of Visual Arts in New York City" (Eisner, 1985).

Like McCloud's work, Eisner's three textbooks provide a visual mix of history, instruction, comic examples and instruction. At the very beginning of *Comics and Sequential Art*, he gives a great historic overview of modern comics and why they are so powerful.

In modern times daily newspaper strips, comic books and more recently, graphic novels provide the major outlets for sequential art. For many decades the strips and comics books were printed on low-grade newsprint never intended for a long shelf life. The often-ancient presses utilized for printing comic books and Sunday comic strips could not even guarantee proper color registration or clarity of line. As the form's potential has become more apparent, better quality and more expensive production has become more commonplace. This, in turn, has resulted in slick full-color publications that appeal to a more sophisticated audience, while black-and-white comic books printed on good paper have found their own constituency. Comics continue to grow as a valid form of reading.

The first comic books (circa 1934) generally contained a random collection of short features. Now, after nearly three quarters of a century, the routine appearance of complete "graphic novels" has, more than anything else, brought into focus the parameters of their structure. When one examines a comic book feature as a whole, the deployment of its unique elements takes on the characteristic of a language. The vocabulary of sequential art has been in continuous development in America. From the first appearance of comic strips in the daily press at the turn of the twentieth century, this popular reading form found a wide audience and in particular was a part of the early literary diet of most young people. Comics communicate in a "language" that relies on a visual experience common to both creator and audience. Modern readers can be expected to have an easy understanding of the image-word mix and the traditional deciphering of text. Comics can be "read" in a wider sense than that term is commonly applied. (Eisner, 1985)

Comics as a medium of expression are a mix not only between *images* and *words* but also there is an element of *time*. This is one of the key points that both McCloud and Eisner are trying to make. By placing these images and words next to each other in a sequence, it creates an effect on the reader that is quite unique. The *art* and *power* of comics lies in the unique effect. A little later in the chapter, we will talk to Chris Warner about the effect and we will also have some specific examples of how you can use this for your SF prototype. But before we get into the specific details, let us look at the Silver Age of Comics—a time when many people think that science save comics.

HOW SCIENCE SAVED COMIC BOOKS

By 1955, comic books were in a terrible state. The popular comic books of the early 1950s were titles like *Crime SuspenStories, Shock SuspenStories* as well as horror comics like *Witches Tales, House of Mystery, Strange Tale*, and *Horrific* and *Fanstastic Fears*. These popular books were action packed, with lurid stories of violent murder and revenge. The covers featured severed heads and eyes literally popping out of their sockets. A noted comics authority,

Lawrence Watt-Evans, described it this way: "Horror comics began to show more gore, more violence, ever more explicitly. It became a matter of topping what had come before: if one issue showed a man killing his wife, the next would top that by having him hack his wife to pieces, and the next would top that by having him eat her corpse" (**Goulart, 2001**).

Inevitably, there came a backlash against these comics. In 1954, an American psychiatrist named Dr. Fredric Wertham published a book called *Seduction of the Innocent*. In the book, Wertham rails against the negative effect of comics books on children, citing that they caused juvenile delinquency. Wertham claimed that comics led minors to engage in sex, drug abuse and violence. *Seduction of the Innocent* was a best seller and brought about a public outcry against the comic book industry. It got so bad that the U.S. Congress launched a Congressional inquiry into the allegations, brining about the Comics Code Authority—a way for the comics industry to voluntary self-police itself.

Comics suffered. Two men were uniquely responsible for the recovery of the industry, super-hero comic books and brining about what is known as the Silver Age of comics. The two worked at the biggest rivals in the industry: DC and Marvel.

> *[Julius] Schwartz, a well-known science fiction fan and literary agent, had been working as an editor for DC since the 1940s. In early 1956, he was given the job of reviving interest in DC superheroes. His vehicle was a new comic, titled Showcase, which featured try out stories for new superheroes. If a character sold well in its Showcase appearances, it was given its own comic. If sales were poor, the character was dropped.* (**Gresh and Weinberg, 2005**)

Schwartz took a nearly forgotten superhero from the 1940s called The Flash and completely revamped him. Schwartz felt that readers wanted their superheroes to be more authentic. They wanted some scientific backing for the heroes' amazing superpowers. Even if the science was farfetched and a little thin in places, Schwartz believed this would make the characters seem more real and make them more popular.

The Flash's superpower is pretty simple: super speed. The original 1940s Flash was Jay Garrick a college student who had gotten his superpowers by knocking over a beaker containing the gas from "hard water." The new flash was Barry Allen, police scientist, who got his powers when lightning struck a rack of chemicals in his lab, covering him with an electrified solution. Barry adopts the identity of The Flash and starts fighting crime.

The first Flash story, *The Brave and the Bold* appeared in *Showcase* and was a huge hit. In 1959, The Flash was given his own comic book and the characters popularity skyrocketed. The comic book continued into the 1990s and The Flash was popular enough at one time that he got his own TV series. That first issue of Showcase from 1959 is considered by many to be the beginning of the new Silver Age of comic books. Lasting until the 1970s, the Silver Age saw comics reach new heights of popularity and creative expression.

But it was not just Schwartz's mildly more scientific explanation of The Flash's origins that made it so popular. Regularly, in the comic, if The Flash was showing some new ability or amazing feat of speed, the panel would include an asterisk (*) in the scene description, followed by an editor's note. The note would attempt to give a scientific explanation for The Flash's super-feat.

In 1961, The Flash battled The Trickster in a comic titled "The Trickster Strikes Back!" (Flash #121). Early in the story, the "sultan of speed" foils a couple of thieves as they steal a jeweled necklace from an apartment. Once "the scarlet speedster" captures the hoods, he takes them to the police station, but before The Flash can turn over the criminals and the necklace, the Trickster snatches the necklace from our hero's hands with an extendable hook! "In a twinkling the world's fastest human has deposited his two captives in the police station and …" The Flash races up the side of the police stations thinking, "The trickster won't escape me this time! His daring is going to spell his downfall if I have anything to say about it !"

As you can imagine, the panel shows The Flash racing straight up the side of the police station in a blur of speed. At the bottom of the panel there is: "Editor's Note: Flash's astounding speed enables him at times to defy gravity itself !"

This formula was such a success that Schwartz would go one to create or revamp a whole stable of superhero characters, all with more scientific reality built into the comics. Some of the most-remembered characters of his were Green Lantern, Hawkman, Justice League of America and the Atom.

Some of the scientific explanations for the superheroes' powers got so long that they could not fit in a footnote at the bottom of the panel. In 1961, Schwartz resurrected the Atom and turned him into Ray Palmer, graduate student and research physicist. In this revamped story, *Birth of the Atom*, Ray finds a piece of a white dwarf star that has smashed into the Earth. Taking it back to his lab, where he is working on a reducing ray, Ray turns the material from the star into a lens and finds he has the ability to shrink any object to an impossibly small size. Later, in the first book, Ray and his girlfriend Jean and a group of school kids get trapped in a cave. They only way out is for Ray to secretly shrink himself and go for help. The Atom is born!

In the next story, *The Battle of the Tiny Titans*, Ray has made himself a tiny suit and is now ready to protect the world when a "mysterious way of seemingly impossible crimes strike Ivytown" (Showcase #34). In their 2002 book, *The Science of Superheroes*, Gresh and Weinberg relate the action:

In the course of his battle with an alien space traveler, Ray performed for the first time what was to be his most unusual stunt as the Atom. He dialed the villain's phone number, turned on a metronome, shrank to microscopic size, and then, when the phone was answered by the crook, traveled through the phone line to the other receiver! The amazing feat couldn't be explained in the panels of the comic, so a special footnote at the end of the issue gave the readers the details. (**Gresh and Weinberg, 2002**)

At the back of the comic *Showcase #34* in a section called *Inside the Atom*, the writers explain the science:

AND NOW for the explanation of the amazing telephone trick pulled by The Atom in "The Battle of the Tiny Titans":

When he dialed the telephone, The Atom caused an electrical impulse to travel from the telephone to the central exchange where, by means of an "electric brain," these impulses were released to travel on to three conductors (these conductors select the telephone dialed, either causing it to ring or send back a busy signal).

When the call was completed and Carl Ballard [the villain] lifted the phone, The Atom leaped into the transmitter, merging the atoms of his microscopically small body with those of the thin diaphragm which is part of the transmitter. This metal diaphragm is set in vibration by sound waves—either the human voice, or as in the case of The Atom, by the sound of the metronome. In vibration, the diaphragm pushes against the carbon granules which are also a part of the transmitter, crowding them together and permitting electric current (which is caused by the sound waves) to travel almost instantaneously at the telephone at the other end of the call.

The Atom was hurled along at telephonic speed by these impulses which (at the other telephone) were translated back into sound waves by the receiver. (Gardner, 1961)

This scientific explanation is questionable at best but interesting because of the lengths that Schwartz, his writers and artists went to so that they could provide explanations to their readers. The readers loved it. Whether it was adding more science to the origins of characters like The Green Lantern and The Flash or adding in detailed and elaborate facts, there was something about the marriage of science and comics that people found exciting. A little bit later in the chapter, we will talk to Chris Warner about why he thinks comics and science go so well together, but before we do that, let us take a look at the other innovator of comics Silver Age: Stan Lee.

I will let Gersh and Weinberg introduce us to Stan the Man:

> Meanwhile, at much smaller Marvel Comics, Stan Lee, another longtime veteran of the comic business, was told by his publisher to create a team of superheroes to match the popularity of DC's newest sensation, the Justice League. [created by Schwartz] Lee invented a group named The Fantastic Four. Soon after, he came up with another superheroes comic, the Incredible Hulk. And within a year, he added The Amazing Spider-man to the Marvel roster. However, Lee didn't follow Schwartz's model of making heroes scientifically plausible. Instead he tried another new idea to comics. He made them into soap operas … For the next ten years, Marvel and DC offered readers a distinct choice. Marvel had the continuing soap opera adventures, aimed at teenagers and preteens, and filled with angst, emotions and melodrama. DC pushed superheroes with a much more science fiction look and air of plausibility in even their most impossible tales. (Gresh and Weinberg, 2005)

To get a more informed perspective, I sat down with Chris Warner to have a conversation about the history of comics, their relationship to science and the craft behind the art.

A CONVERSATION WITH CHRIS WARNER

Chris Warner broke into comics professionally as a penciller for Marvel Comics in 1984. He worked on *Alien Legion*, *Moon Knight*, and *Dr. Strange*, then joined up with Mike Richardson and Randy Stradley when they began Dark Horse Comics in 1986. Today, Dark Horse is the largest American comic book and Managa publisher, based in Milwaukie, Oregon.

Warner drew the cover and story, "Black Cross," for the very first Dark Horse title, *Dark Horse Presents* #1. He has drawn and written a bunch of titles for Dark Horse over the years like *The American, Predator, Aliens vs. Predator, The Terminator, X, Barb Wire*, and eventually moved full-time into editorial in 1998.

Warner grew up in Oregon and his office at Dark Horse is in the old five-and-dime where he used to buy comics and science fiction paperbacks when he was a kid. To start off our conversation, I asked Warner about how he got into comics in the first place.

CHRIS WARNER

My first real memory of reading comics was DC's Batman. I liked the fact that he was a detective. I never liked Superman very much. I always wondered how heroic is somebody really if they're invulnerable and super-strong? But Batman had to use his wits and his fists to survive. Plus just the fact that he was Bat, which when I was a kid, I thought was creepy and cool.

When I was a kid, comics were the only place where you could find those kinds of stories. Comics seemed so weird and compelling to me because they were this small group of characters that all lived in the same world together and all the books were kind of related. That was fascinating. For as simple as comics were there seemed like there was always a gigantically elaborate architecture behind the stories and that fascinated me.

* * *

Starting in the Silver Age, the bookends of the comic industry were Marvel and DC. Marvel had Spiderman while DC had Superman. Through the 1950s, 1960s and 1970s, both Marvel and DC differentiated themselves not only with their characters and super heroes but also by how they wrote and drew the comics.

CHRIS WARNER

DC Comics' stories always seemed very external to me. The heroes were fighting crime and saving the world but seemed to be just going about their jobs. Marvel Comics had this almost mythological framework. Marvel stories were about these clashes on a gigantic scale, but the stories seemed just as titanic, on a personal level, to the characters. DC stories seemed less personal, less intense.

The Marvel stories were character driven. Some people described them as superhero soap operas, which is just shorthand for saying the characters had inner lives. The mainstream characters of the day seemed to be all about the hero business, saving this, taking care of that, protecting their secret identities. They had their day jobs and their love interests, but the stories lacked any real emotional content. In Marvel comics, every page seemed to carry a sense of urgency. Even having great powers seemed as much a curse as a blessing. That really made me want to read them over and over again because they were just so intense.

Stan Lee and especially Jack Kirby from Marvel were heavily influenced by classic mythology. In a way, Kirby and Lee created a modern urban mythology. Marvel comics operated on a grand scale. Even the artists were different. Most comic artists of the day tended toward an accepted commercial realism. They worked more in a classic newspaper-strip influenced style. Marvel artists were more exaggerated and were viewed negatively by a lot of mainstream professionals. Some thought their work was ugly and crude, but to me it was so much more powerful. There never really seemed to me to be any particular reason to maintain a visual style that was married to reality. Why not push it? If you're telling stories within an exaggerated, bigger-than-life narrative framework—superheroes, monsters, aliens—why not push the visuals and make them bigger than life, too?

HOW TO TELL A COMIC BOOK STORY

Although comics are a hugely popular medium, few people know how to actually go about creating them. I asked Warner what he thought was unique about comic book storytelling and what advice would he give to writers.

CHRIS WARNER

Comics are designed to present information clearly. Comics work with a limited number of images and words to tell a story and make a series of static images come to life in the imagination of the reader, so the reader must understand visually, at a glance, what's happening in each picture. You have to know instantly what the picture is telling you, whether it's a specific action or the emotional states of the characters. That sounds simple, but it isn't.

I tell comic artists this all the time: there are things in a picture the reader needs to see but, more importantly, there are things the reader needs to look at. What comics strive to do in each panel is to focus the reader's attention on the important elements, to make the reader focus on them while seeing the peripheral stuff without looking at it. Two characters are having an animated conversation while walking though a park. We need to see the park, but we likely don't need to look at it. We should be focusing on the characters. In this sense, comic illustration is very different from traditional illustration, which is often designed and facilitated to make the viewer's eye linger on the image. If I'm trying to sell you something, the longer you stay focused on the image, the better chance I have to sell you the product.

But when creating comics, the goal is to move the reader along. You want the reader to see the picture, read the text, move to the next panel, read the text, and keep the story flowing. If the reader's attention wanders and focus is pulled by unnecessary "eye candy," the artist may have succeeded in drawing pretty pictures but failed in visual storytelling. The instant a reader's eye begins to wander, or when the reader doesn't immediately absorb the picture's message, the story spell is broken.

When a reader understands everything at a glance, the story captures the reader, who can reach that blissful stage where they aren't aware that they're actually sitting in a chair with a pamphlet in their hands. They're just absorbed in a story.

When you compose a panel, the most important thing is the most important thing. In every picture there's a key visual message, and you really have to concentrate on how to arrange and treat the individual elements to get that message across effectively.

The same thing is true with the text. You don't have a lot of space to tell your story. A typical comic book of twenty-two pages has perhaps a hundred-odd drawings. Most of what happens in the story, when determined in real time, happens between the panels. Each panel is just a snapshot of time. The reader's mind is filling in the gaps. It's like you're leaving a trail of crumbs to lead the reader through the story. When the key images and text hit the right spots, the reader seamlessly fills in the gaps, follows the trail. If the reader misses those keys, the whole structure collapses, and the reader loses the trail. You have to boil down the story to those key snapshots and key text, and what you leave out is just as important as what you put in.

It's like trying to hold up a heavy object with straws. If you have enough straws it's easy. But, if I give you a limited number of straws, you have to be really careful about what you're going to do to hold that thing up. That's what we're doing with comics. You're trying to find a way to tell a very rich and complex story, while being able to just hit the high notes.

It's important for the writer to really understand what's important and to understand the key elements the reader needs to stay absorbed in the narrative. The writer has to know what the story is really about. That sounds obvious, but it's actually quite easy to not know what the story is about, even if you're writing it. It's easy to get caught up in the plot, and the characters, and the genre trappings and let the point of the story get past you. There has to be a point to the story.

* * *

It has been said that comics are about extremes. The stories are about dire situations and the end of the world. The characters are superheroes and supervillains. The action is always over the top and like Warner explained comics have to get this all across with as few pictures and words as possible. But comics are also about leaving things out, making use of the spaces between the panels as Mc-Cloud and Eisner talked about earlier in the chapters. I asked Warner about his approach to this type of storytelling.

CHRIS WARNER

You're trying to strip away the unnecessary. The more cluttered and undirected the visual, the less likely the reader is to have that image stick in the memory. You need the images and text to stay with the reader from panel to panel—that's what gives the narrative momentum, when it carries the reader away. It's important that the image stays with you, and the more visual information loaded in the panel, the less likely the reader is to hold on to that image.

In film it's much easier for those key images to stick, because the image is forty feet high in a dark room. What else are you going to look at, the guy's head in front of you? But with comics, distractions are much more possible, and you have to find a way to pull the reader in with at-a-glance clarity. The clear image will more likely stick in the memory.

This is especially important in the case when a comic is serialized. You don't want to make the reader go back and reread the previous month just to remember what's going on. I find I stop reading a comic series when I can't remember what happened in the previous issues. They may have been compelling, but they weren't structured in such a way that is conducive to visual memory. I can still remember Jack Kirby panels from when I was a kid. I remember the entrance of the character Black Panther like it was yesterday, and it's been forty-five years since I first read that comic.

What makes good comics is understanding what's important, winnowing down a story to its critical moments that makes a good comic so powerful. In a comic you have only static images … there's no interpretation of the image needed. It's just there. That's what it is. The words that are on the page are there. You must distill a story into a limited number of words and pictures. Comics' strength rides on precisely this, because with limited pictures and words, the key ideas can be gotten across very clearly with a minimum of narrative distractions.

With prose, you need a lot verbiage to set things up. You have to explain how things look and where they are set. But in a comic, you can take care of that in a visual instant. Here's a picture, that's what it looks like and I don't need to explain that now because now you've seen it.

BDJ

So what makes comics different than prose fiction?

CHRIS WARNER

I think it has to do with the serial nature of comics. It prevents your mind from wandering too much. When you're reading a book and you're reading great writing, and you're suddenly being carried away by the brilliant language, and all of a sudden … where the hell am I in the story?

For example, I love Cormac McCarthy's fiction. Some of his sentences can go on for half a page. He begins describing in painstaking, breathtaking detail the terrain of northern Mexico or southern Texas, but by the end of the passage, he's also somehow described the totality of the human condition and the futility of existence. I don't know how he does it. I have to take a step back and read it again, sometimes several times. But when I do that it breaks the spell of the narrative. It's worth it, but the very brilliance of his language can subvert the momentum of the story.

But when you read comics you're going from panel to panel to panel and your mind is filling in the gaps. You're creating the connections between panels; you're filling in the story. When you read each panel, it's as though you actually experienced what's in the gaps, even though you were shown none of this.

So in essence, there's a cognitive compression there that is different in comics than in film or fiction. You look at a panel and then you look at the next panel and then the next, and you know that there is stuff in between, but you don't really have to consciously think about it.

It's at this point in comics that the reader has become so active and their mind is so engaged that they are probably not going to wander off and lose the point of the story.

That's especially important when you're trying to teach something or get across a complex idea. In these cases, you don't want the reader to get too caught up in peripheral details and implications. There's time for that later. While reading the story, you want the reader to focus on main ideas. Comics can do that better than almost any medium because of their simplicity and directness of pictures and words.

SCIENCE IN THE GUTTERS

The cognitive compression that Warner explained is at the heart of why comics make such a good medium for SF prototyping. Comics by nature must be very precise. They need to be trimmed down, making an efficient use of space, leaving out as much as possible but still getting the point across.

Comics are visual, they show. When you are telling a story with a short story and you want to describe a house, you have to take the time to paint the detail with words. You could do it in a few paragraphs or a few sentences, but to have a specific level of detail it takes time. If you were showing the same house in a short film, that too takes time to set up an establishing shot of the house and possibly a close up to show specific details. But in a comic you would simply draw the house. You could give the house as much detail as you would like, but it can all be accomplished in a single panel, and the reader can take it in very quickly. This is something that comics can do that our other medium for SF prototyping cannot.

Comics also have text, they tell. Comics not only tell the reader a quick line about the action or a snippet of dialogue but they do it quickly. In those panels, there is no room for long sentences or overly detailed descriptions. As Warner said, you have to find out what is the most important thing and then move on.

When you think about comics that way, it is almost as if they are perfectly suited for SF prototyping. The task of pulling together a comic is kind of like an engineering task, it must be precise and exact. You have to get across your idea as succinctly as possible and move on. The reader can fill in the gaps as they move from panel to panel. And it is that filling which will helps us the most with our SF prototype.

Scott McCloud in his book *Understanding Comics: The Invisible Art* described this characteristic of comic storytelling as Blood in the Gutter. Here is the illustration:

In one panel a man run's for his life screaming, "No! No!"

Over his shoulder we see a ax wielding maniac yelling, "Now you die!!"

In the next panel we see a wide shot of the city at night. The moon hangs in the air and across the panel is the sound effect, "EEYAA!!"

McCloud says, "See that space between the panels? That's what comic aficionados have named 'The Gutter.' And despite the unceremonious title, the gutter plays host to much of the magic and mystery that are at the very heart of comics!"

In the next panel, McCloud stands between the two previous panels we saw; the murderer with his victim and the dark night sky with the cries of death.

McCloud explains, "Here in the limbo of the gutter, human imagination takes two separate images and transforms them into a single idea."

After a little more explanation McCloud wraps up saying, "Every act committed to paper by the comics artist is aided and abetted by a silent accomplice. An equal partner in crime known as the reader."

Again McCloud takes up back to the two panels: the murderer with his victim and the dark night sky with the cries of death. "I may have drawn an axe being raised in this example, but I'm not the one who let it drop or decided how hard the blow or who screamed or why. That dear reader, was your special crime, each of you committing it in your own style. All of you participated in the murder. All of you held the axe and chose the spot. To kill a man between panels is to condemn him to a thousand deaths" (**McCloud, 1993**).

Blood in the gutters refers to the fact that most comic book violence happens between the panels. Just as McCloud showed us with his example, the axe wielding murder happened between the panels. It happened in the gutter of the page. When that murder happens in the gutter, we as the reader fill in the details. We naturally imagine the death. The reader connects the dots between what was seen in the first panel and what they see in the next. This is unique to comics and something we can use for our SF prototype.

The title of this chapter is Science in the Gutters because I think we can use the precise nature of comics and the gutters to help us and the reader imagine the implications of our science or technology on people, society and larger complex systems. If we describe the science accurately and succinctly and we follow that with the effect of that science on the world then the reader will fill in the gaps. The actual act of reading the comic, moving from panel to panel will in the reader's mind paint the implications in a way that neither the short story nor short film could. In the case of using comics as an expression of your SF prototype outline, it will be the things that you leave out that are important and every person who reads the comics will paint their own implications.

TURNING YOUR SF PROTOTYPING OUTLINE INTO A COMPELLING COMICS STORY

Will Eisner tackles the question, "What is a story?" in his book *Graphic Storytelling and Visual Narrative*. "All stories have a structure. A story has a beginning, an end and a thread of events laid upon a framework that holds the two together. Whether the medium is text, film or comics, the skeleton is the same. The style and manner of its telling may be influenced by the medium but the story itself abides.

The structure of a story can be diagramed with many variations because it is subject to different patterns between its beginning and end. A structure is useful as a guide to maintaining control of the telling" (**Eisner, 1996**).

Below this description in the book Eisner lays out five simple points for a story:

- Introduction or setting,
- Problem,
- Dealing with problem,
- Solution,
- End.

That's it. It is that simple. Again you will notice the similarities between Eisner's approach and the diagrams we discussed in our two previous chapters. There is something universal about these approaches to storytelling and it is up to you to change and mold it however you see fit.

Peter David is a writer who has tackled not only comics but also TV, movies, video games and novels. His comic book work includes writing for *Spider-Man, X-Factor, Wolverine, Supergirl, Captain Marvel*, and *Hulk*. In 2006, he wrote a book specifically about writing for comic books. It is called, not surprisingly, *Writing for Comics with Peter David*. In the plot and story structure section of the book, David recognizes that the three act structure and the basic story structure we have been examining throughout this book are not entirely new ideas:

> *It's not as if someone recently invented the three-act structure and all new stories adhering to it. Rather the fundamental concept has existed for decades and is simply a sound way of telling a story. Aristotle said that dramas consisted of three parts: a beginning, a middle and an end. That may seem self-evident, but remember, the Greeks invented drama, and what we take for granted now was ground breaking a couple of thousand years ago …*

> *In recent years, Artistole's simple observation of beginning, middle and end has been codified, institutionalized, some might even say mummified … I should emphasize that this is a structure, not a formula. A formula results in a sameness that makes everything seem overly familiar and predictable.* (David, 2006)

David goes on to explore the three-act structure as it is specially applied to comic book storytelling. He breaks down each act with examples from Marvel's *Fantastic Four*, *Spider-Man*, *Hulk*, and Dark Horse's *Spyboy*. If you are interested in getting more detail about the process and the form of your script, I recommend David's book along with McCloud and Eisner.

THE SUPERMAN—ANALYSIS OF COMIC AS AN SF PROTOTYPE

I realize that I have spent this entire chapter talking about the visual medium of comics books and not once shown you a single panel from a comic. Let us change that. We are going to take a look at *The Supermen*, a six-page comic that was original published in Out of this World #3 in 1957. The comic is by another legend in the comics industry Steve Ditko. Ditko is best known as the co-creator of Marvel's Super-Man and Doctor Strange. He worked closely with Stan Lee during the Silver Age of comics. Ditko was inducted into the comics industry's Jack Kirby Hall of Fame in 1990, and into the Will Eisner Award Hall of Fame in 1994.

After you have a look at the story once, try giving it a few more read-throughs. The first time just read it like a story. When you go back and look at it think about what Chris Warner said earlier about understanding what is important and only showing that. Have a look at what has been left out and what happen in the gutters.

After that we will reverse engineer the story, breaking down the story into the five easy steps of our SF prototyping process.

OUT OF THIS WORLD

THEY WERE NO LONGER HUMAN! AND YET EVEN THEN THEY WENT ON! BUT THERE ARE SOME THINGS MAN MUST NOT MEDDLE WITH -- I KNOW! I SAW WHAT HAPPENED TO....

The SUPERMEN

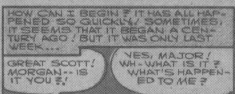

HOW CAN I BEGIN? IT HAS ALL HAPPENED SO QUICKLY! SOMETIMES IT SEEMS THAT IT BEGAN A CENTURY AGO! BUT IT WAS ONLY LAST WEEK...

GREAT SCOTT! MORGAN -- IS IT YOU?!

YES, MAJOR! WH-WHAT IS IT? WHAT'S HAPPENED TO ME?

I WAS WORKING NEAR THE ATOMIC REACTOR AND SUDDENLY THIS HAPPENED! I-I FEEL AS IF I'M GOING INSANE!

STEADY, YOU'RE FAR FROM IN-SANE! BUT AS TO WHAT'S MAKING YOU GLOW-- I DON'T KNOW! SOMEONE GET DOCTOR LEWIS!

OUT OF THIS WORLD

MORGAN WAS THE FIRST! HE STUMBLED INTO THE CONTROL ROOM, AND...

WELL, DOCTOR?

THE GLOW IS CAUSED BY TOO MUCH EXPOSURE TO ATOMIC RADIATION -- BUT I DON'T SEE HOW OR WHY?

MAJOR, AS A DOCTOR, I'VE BEEN AGAINST THIS EXPERIMENT ALL ALONG! HERE WE ARE, ISOLATED IN THE ROCKIES-- WITH ENOUGH URANIUM TO BUILD A HUNDRED A-BOMBS...

THAT'S OUR JOB, DOCTOR! WE ARE SOLDIERS, EVEN IF YOU'RE NOT!

I KNOW! YOU'RE HERE TO EXPERIMENT, TO LEARN WHAT HAPPENS TO THE HUMAN MIND WHEN IT IS EXPOSED TO CONSTANT RADIATION...

BUT ONLY TO CONTROLLED RADIATION, MAJOR, THERE MUST BE A RADIATION LEAK!

THE DOCTOR IS RIGHT! WE'VE GOT TO GET OUT OF HERE!

THE DOCTOR IS AGAINST THIS EXPERIMENT, BUT HE'S A MAN OF SCIENCE! HE VOLUNTEERED! YOU DON'T HAVE THAT CHOICE-- YOU'RE A SOLDIER!

IF THERE'S A LEAK -- WE'LL FIND IT! BUT WE STAY! THE DOCTOR'LL TAKE CARE OF YOU! ONLY DON'T LET ME HEAR ANY MORE TALK OF ABANDONING OUR JOB!

I REMEMBER SO WELL, THAT FIRST DAY-- EVEN THE MAJOR WAS FRIGHTENED! BUT HE HID IT! AFTERWARD, HE LED THE SEARCH PARTY...

ANY LUCK, MAJOR!

NONE! THE WHOLE AREA IS HOT! WE KNOW THAT! THERE'S A CERTAIN AMOUNT OF RADIATION EVERYWHERE! BUT WE CAN'T FIND A LEAK!

I SEE! I'VE GIVEN MORGAN A SEDATIVE AND PUT HIM TO BED! HE SEEMS HEALTHY ENOUGH, BUT... MAJOR! I'VE GOT AN IDEA!

ABOUT WHAT'S WRONG WITH MORGAN?

OUT OF THIS WORLD

YES, I STILL SAY WE SHOULD LEAVE THIS PLACE! I'VE EXAMINED MORGAN! HE'S... CHANGING! HIS SKULL, IT'S LARGER! I... I THINK IT'S GROWING!

GROWING? THAT'S INSANE! I'LL HAVE A LOOK!

ONLY-- IT WASN'T INSANE! WE ALL WENT! AND THE DOCTOR WAS RIGHT...

YOU'RE RIGHT! IT'S FANTASTIC!

IF ONLY THE MAJOR HAD TAKEN US AWAY THEN! BUT HE DIDN'T! HE WAS LIKE STEEL! SO, IN THE MORNING...

STEVENS! IT'S HIT STEVENS, TOO!

STEVENS, JOHNSON--IN A FEW HOURS, IT WAS TOO LATE! BECAUSE IN A FEW HOURS, THERE WAS NO LONGER A CHANCE THAT THE MAJOR WOULD LEAD US TO SAFETY...

SO, IT'S HAPPENED TO YOU, TOO! I'VE JUST BEEN TALKING TO MORGAN, MAJOR! I KNOW WHAT'S HAPPENING TO US! MORGAN HAS BECOME TELEPATHIC! HE CAN READ MY MIND!

WE'RE... CHANGING! RAPIDLY! SOMEHOW, THE RADIATION HAS SPEEDED UP THE PROCESS! OUR BRAINS ARE GROWING BY LEAPS AND BOUNDS!

THEN IF IT KEEPS UP, WE'LL BE-COME -- SUPERMEN!

IT'S INCREDIBLE! BUT SOMEHOW I CAN'T HELP... LOOKING FORWARD TO IT! I WONDER WHAT IT WOULD BE LIKE!

I THOUGHT YOU'D FEEL THAT WAY! SOMEHOW, NONE OF US WANT TO GET AWAY ANY LONGER! ALL WE HAVE TO DO NOW, IS... WAIT!

3

OUT OF THIS WORLD

DO YOU UNDERSTAND? HUMAN EMOTIONS AND FEARS DROPPED AWAY FROM THE MEN./ IN A FEW DAYS, THEY DIDN'T EVEN SPEAK, TO EACH OTHER...

WHAT IS THE HUMAN BRAIN, ONLY A MACHINE? AND IS THAT WHAT THE MEN BECAME? MACHINES? THINKING MACHINES...?

FOUR DAYS, FIVE--AND THEN THE LAST ACT BEGAN./ AND AGAIN, MORGAN WAS THE FIRST...

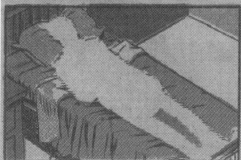

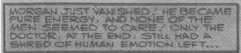

FIVE EASY STEPS—BREAKING DOWN *THE SUPERMEN* AS AN SF PROTOTYPE

Let us look at Ditko's *The Supermen* as if it were an SF prototype. Before we do this, we must remember what the world was like in

1957 when Ditko wrote the piece. In 1957, we were locked in the midst of the Cold War and people were worried about atomic energy. Not much was known about the effects it would have on people and Ditko plays off of this.

1. Set Up:

- The military (the major and his pet monkey) and a group of scientists (Morgan, the doctor, etc.) are working on a uranium experiment at a remote and isolated laboratory in the Rockies.
- The experiment is to study what happens to the human mind under constant radiation.
- The entire test site has a low level of radiation.

2. Inflection Point (Science):

- Dr. Morgan, one of the scientists, is working near an Atomic Reactor and gets a dose of Uranium causing him strange side effects.

3. Ramifications of #2:

- Morgan's side effects: he glows, his skull begins to grow in size.
- As the radiation affects all the men at the lab.
- Morgan develops telepathic powers.
- The longer the men stay at the lab, the stronger their mental abilities.
- They stop talking to each other.

4. Inflection Point (Human):

- Ultimately, Morgan turns into pure energy and disappears.

5. Ramifications of #4:

- Once Morgan had disappeared the men must make a decision to leave and return to being human or to stay, remaining supermen but ultimately disappearing.
- Only the Major's Monkey remains to tell the tale, as he too has become super intelligent and will be converted to pure energy as well.

SILLY SCIENCE BUT A GOOD IDEA

The science of Dikto's story may not hold up to scrutiny and the super intelligent monkey does work as a fun twist ending but we can also see *The Supermen* as a highly affective SF prototype. The story takes a realistic setting: A remote lab in the Rockies, staffed by doctors and the military (we will overlook the monkey).

Next, we introduce a piece of science or a scientific question: what happens to the human brain when exposed to a lever of radiation over a long period of time?

Next: We see the effects. In the story, the men begin to glow, their intellect and heads grow in size and they lose all of their emotions.

Finally: the effect of the science (Low Level Uranium radiation) is that these men lose their will to live and trade their lives to be supermen. It's not a bad story. In six, relatively, simple pages, we explore the implications of Uranium radiation on lab workers. The form of the comic allows for a minimal text and pictures. The effect of the radiation on the men (and the monkey) is left to the gutters, keeping it a mystery and letting the reader imagine and ponder what is really happening to them.

The Supermen provides us with a nice stating off point for considering how you apply your SF prototype outline to the comics medium. We have seen how the form of comics storytelling is unique, providing you with different options when you consider what form you would like your SF prototype to take. Is your idea best suited from comics? What will your science be like in the gutters of your story?

Now it is up to you. Get started on your SF prototype and make sure you have fun.

Excelsior!

. . . .

CHAPTER 7

Making the Future: Now That You Have Developed Your SF Prototype, What's Next?

As we mentioned in Chapter 1, probably the most public and popular example of SF prototypes recently was *Uber Morgen* or *The Tomorrow Project* sponsored together by the Intel corporation (http://newsroom.intel.com/docs/DOC-1490). It was a short story anthology published in Germany in 2010. On the web site, we describe it this way:

"The Tomorrow-Project" is a unique literary project which shows the important effects that contemporary research will have on our future and the relevance that this research has for each of us. Research currently being conducted by Intel in the fields of photonics, robotics, telematics, dynamic physical rendering and intelligent sensors served as the basis to inspire four bestselling authors. The results are four short stories which paint amusing, thought-provoking and hopeful pictures of our future.

Last Day of Work—by Douglas Rushkoff
Today is Dr. Leon Spiegel's last day of work. But he's not just another retiring technology worker: he is the last man ever to work.

Having delayed the inevitable for longer than he should, Spiegel recounts the events that have led to a world where no companies, no money, and no need for employment exist. In doing so, he reveals how humanity nearly allowed technology to bring life to a close, before stumbling upon the truth of man's own culpability for his dire condition. And now that humankind has avoided its dark fate and transcended the previously limited definition of what it means to be human, Spiegel is having a hard time letting go and joining the rest of the world.

The Mercy Dash—by Ray Hammond

In the year 2125, two young Mannheim residents are forced to undertake a high speed drive to save a life. Billy Becker — a successful furniture designer — and his girlfriend Sophie, a medical student, learn that Sophie's mother Hélène has suffered a serious back injury whilst waterskiing. Hélène has a rare blood antibody which means it is unsafe for her to receive ordinary blood transfusions. Sophie also carries the rare blood antibody and she alone can provide blood.

But, in 2125, all major autoroutes through Europe are under networked traffic management and Billy and Sophie are forced to override the automated controls and speed limits in order to get to from Mannheim to Nice before surgery on Hélène is completed.

Billy and Sophie are not alone on their journey. Billy has recently acquired a new virtual assistant, also called "Sophie," who is able to communicate in a very lifelike way; so lifelike that real Sophie starts to get jealous about her rival for Billy's attention …

The Drop—by Scarlett Thomas

Agnes is 32. She lives in a seaside town in Britain about six years (or so) in the future. When she is out running one evening on the seafront she ends up racing a boat and winning. When one of the rowers sends her a message on her GSRcx (a sports watch that includes galvanic skin response data and information on air pollution and wind-speed) she is quite surprised: you can't get messages on this kind sports watch. Everyone uses their 'Box' for communication. It's not clear how she can send him a reply. She doesn't even know which one of the four people on the boat he was, but hopes he's the one with the dark curly hair and the green top. When she discovers that he sent her the message using a mind-control patch, she must learn mind-control in order to send one back.

Agnes' family makes money by generating electricity on exercise machines. They get briefly excited when they get some 'hits' from people watching them in their everyday lives (like reality TV, but live and accessible direct from home-to-home via speed-of-light data transfer), but the hits soon evaporate. It doesn't really matter — everyone, watches the Takahashi family in Tokyo anyway. Agnes' father spends all his time in the (virtual) mountains. Danny, Agnes' younger brother, is obsessed with watching cars on the 'network': a system many people think is very beautiful and even mysterious. Cars drive themselves, using the most efficient route. And all cars have their color chosen by the network. The cars make patterns that can be seen from space (and that anyone can tune in to).

The Drop imagines a future in which telematics, photonics and intelligent devices have changed the ways in which people interact with their world. The world, and the people in it, remain (lovably) flawed, mysterious and sometimes funny—but the technology makes many things so much easier and more interesting.

The Blink of an Eye—by Markus Heitz
A brave new world, thanks to sensor technology and AI! Every movement is monitored. Alexin can walk around inside his house without ever having to flip a switch. The house detects, interprets and makes life easier: doors, lights, electrical and electronic devices, even the toilet seat lifts and lowers automatically. But Alexin knows this is making him dependent—and it's becoming painfully clear to him that AI also has a dangerous disadvantage....

A brief moment in the future and an appeal to use our minds: that's this contribution by author Markus Heitz.

The aim of the *Uber Morgen* anthology of four stories and also the goal broader Tomorrow Project worldwide is to begin conversations with people about the future. Starting this conversation with people about the future they want to live in is something SF prototypes can do really well. Capturing people's imaginations with science fiction based on real technology allows us to have a broader conversation about the kind of future we want to build. But SF prototypes can do more than just start a conversation.

FROM FACT TO FICTION TO FACT ONCE AGAIN: AN SF PROTOTYPE USED IN AI DEVELOPMENT

SF prototypes can also be used in technology and product development. To illustrate this, let us take a look at the work that we have been doing on AI for domestic robots in complex environments. The SF prototype we used as an example in Chapter 2 *Nebulous Mechanisms* was the first in a series of stories that we have been using to develop theories of AI at the University of Essex in the UK and at Monash University in Kuala Lumpur, Malaysia. Each of the SF prototypes in the Dr. Simon Egerton series of stories allows the students, professors, computer scientists and engineers to explore the implications of their work on people, society and broader systems. At the moment, we are about halfway through series and the results have been fascinating.

The second SF prototype in the Dr. Simon Egerton Stories was called *Brain Machines*. If you remember the first story, *Nebulous Mechanisms*, was an exploration of AI and robots that could make both rational and irrational decisions to help them survive in complex environments. *Brain Machines* takes this idea a step further, exploring the idea of free will. First, it examines the concept of free will as it applies to humans. It turns out that it is hotly debated if we humans even have free will. New neuroscience makes this a hotly debated topic. Second, the SF prototype explores how free will might be beneficial to AIs and robots; a scary notion for some and an intellectually challenging problem for others.

Now, as you already know from reading the other chapters in this book, *Brain Machines* is based on a collection of scientific papers and current thinking. We will get into that in just a minute.

Below is a synopsis of the story. If you want the full SF prototype, you will find it in Appendix A.

BRAIN MACHINES
Story: Brian David Johnson
Illustrations: Winkstink

"Jimmy, fix me another drink," Dr. Simon Egerton said as he sat in his cramped apartment buried in the clog of stations that ringed the Earth.

"No problem," Jimmy replied cheerfully and set off for the make-shift bar. Jimmy was a pet project of Egerton's. He was an off-the-shelf clean room assembly bot that Egerton had modified into a somewhat old fashioned service bot. Egerton had been experimenting with Jimmy during his free evenings. At a little over three feet, Jimmy was a cute little guy. His rounded hip joints and oversize half-skull made him teeter when he moved across the floor. He looked like a child just learning to walk.

"How are you feeling?" Egerton asked Jimmy.

"Fine thanks," he replied, mixing the gin and tonic. "We're running low on Tanqueray." He turned and waddled a few steps with the drink, concentrated and careful not to spill.

"Thanks." Egerton took the drink and searched the bot for anything out of the ordinary.

"No problem at all." Jimmy waddled back to the make-shift bar and tidied up.

Egerton sat his fresh drink on the floor next to the chair, lined up with eight other untouched cocktails. After a moment he asked, "Jimmy, will you fix me a drink?"

"No problem," Jimmy replied cheerfully and started on the tenth drink.

Egerton puzzled at the back of the little bot. He knew that Jimmy knew he wasn't drinking the gins. Egerton knew that Jimmy knew he was being tested and that it was silly to keep on making gin after gin. But Jimmy wouldn't react. He wouldn't break out of his service duties and ask what was going on. Why wasn't Egerton drinking the cocktails? Was there something wrong with them? It was a problem of will: free will. Jimmy had all the capabilities to question what was going on but he wouldn't do it. It was a problem Egerton had been trying to crack for over 6 months.

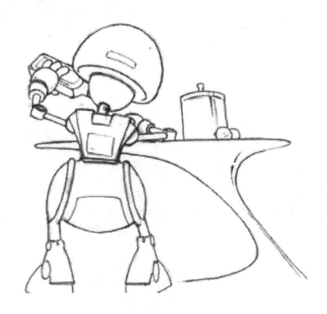

"We're running low on Tanqueray," Jimmy said finishing the drink.

Egerton's phone rang. "Simon Egerton." He leaned into the phone, weary of any call that got to him with such little information attached. The caller's ID was: Ashley Wenzel.

"Simon? Simon, can you hear me? I cannot tell if this thing is working?"

Egerton recognized the caller immediately. The pinched and impatient face of Dr. Sellings Freeman came over the cheap phone connection. Sellings had been Egerton's professor and sometimes mentor at university. He was neither a good professor nor a good mentor but Egerton had learned a lot from the pompous old man. Plus Sellings had gotten him his first research grant so Egerton felt forever in his debt.

"Hello Sellings," Egerton replied. "Why does your name come up as Ashley Wenzel?"

"I had to borrow this girl's phone. I cannot explain. I do not have time. I think I am breaking some law by even making this call." Sellings was distracted and tense. "I need your help Simon. I need your help and there is no one else I can ask."

"What's wrong?" Egerton grew concerned. He'd never seen Sellings upset; the pompous old man was unflappable. "What's happening?"

"I can't say right now … what? Yes, wait just one moment" Sellings' attention shifted. "I'm almost done."

"I need my phone back," a girl's voice came over the line. "You said …"

"I am almost finished," Sellings answered annoyed and frustrated. "Simon. I need you to come to Maralinga Gardens right away. There's no one else I can ask to do this …"

"You mean you're on Earth?" Simon was shocked. "Why are you down *there*? What could you possibly …"

"1370 Anangu Way," Sellings barked. "That's where I am staying. 1370 Anangu Way, Maralinga Gardens. Can you come right away?"

"I can try," Egerton wasn't sure what to say. "There's some things I need to …"

"I told you I was almost finished …"

The connection went dead.

Egerton watched the dead phone for a moment, wondering what to do when he remembered Jimmy. "Oh I'm sorry Jimmy," he said, seeing the little bot standing in the middle of the floor trying the keep the perspiration on the glass from running onto the floor.

"We're almost out of Tanqueray," Jimmy said happily handing over the cocktail.

"I'll go get more." Egerton set the glass next to the chair with the others and stood up. Grabbing the Tanqueray bottle, he slipped into the kitchen nook and refilled it with water. Stepping back into the living room he handed the bottle to Jimmy, who checked it, wiped it with a cloth and returned it to the bar.

"Jimmy, will you fix me a drink?"

"No problem."

Egerton stared at Jimmy's half-skull and wondered what to do.

Note: Dr. Egerton then makes the dangerous trip to Maralinga Gardens in Australia to help out his old professor Dr. Sellings Freeman. Once there Egerton discovers that Sellings is embroiled in a legal and scientific hot bed of debate that has galvanized around a murder trial. An ex-assistant of Sellings has committed a double murder and is claiming that he was not responsible for the crime because he doesn't have free will. This is problematic because Selling's research has proved this to be true, that scientifically humans do not have free will. The only problem, aside from the media feeding frenzy, is that Sellings has also learned that when the general public questions their own free will they develop a kind of mental virus that drives them crazy.

I won't give away the ending but ultimately Egerton and Sellings figure out a solution and Maralingua Gardens is saved. But there is still the problem of the little robot Jimmy and free will. The SF prototype ends like this:

"Jimmy will you fix me and Dr. Freeman a drink?" Egerton asked.

"Sure thing," the bot replied cheerfully. "As I remember Dr. Freeman likes Traditional Pimms number one …"

"That would be lovely Jimmy," Sellings replied with an astonished smile. "Thank you for remembering."

"No problem at all," Jimmy answered then set about making the drinks.

Egerton and Sellings sat in Egerton's cramp apartment in the clog of stations that ringed Earth.

"Simon, he really is a silly robot," Sellings said watching Jimmy's small child-like body work at the bar.

"I know," Egerton grinned, pleased Jimmy. "He's a funny little guy but he keeps me company …" he trailed off then added, "… he's the first one, you know. Jimmy will do down in history."

"Yes, yes, yes … I read your paper and saw your telecast," Sellings replied, unable to hide his annoyance. "*Teaching Free Will as an Alternative to Metathought Safeguards* … you couldn't have thought of a catchier title?"

"I hope you're not too mad at me." Egerton was sheepish. "I had to leave Maralinga. I didn't want to leave you like that but I couldn't lie and when I finally figured it out I just couldn't …"

"I understand," Selling didn't hide his dissatisfaction. "I was more confused when I heard. Bridgette was livid."

"I saw Bok got life in prison," Egerton said, trying to move on.

"Yes," Sellings sighed. "My lie locked up a murderer, ended my career and saved Maralinga Gardens."

"Your career isn't over …"

"Stop," Sellings waved his hand in front of his face. "Let an old man be bitter if I want to. Yes. Yes. I still have work but it's not the same."

"But seriously Sellings," Egerton had rehearsed the next words for several weeks. "I do have you to thank for helping me. If I hadn't come to Maralinga …"

"Shut up Simon!" Sellings interrupted. "I don't care about your speech. Just tell me what you did." He pointed at Jimmy. "What did you do?"

"I just told him he was free," Egerton laughed. "That's it. It was simple. I taught him he had free will. That's it really. Then everything changed."

Sellings sucked in air with astonishment. "That's it? How funny…."

There was a gentle pause in the conversation as both men played around with their own thoughts.

"There's just one thing," Egerton interrupted as Jimmy returned with the drinks.

Sellings took a sip of his Pimms and replied, "What's that?"

"Well …," Egerton paused. "I think Jimmy is developing a soul."

"Really?" Sellings stared at the cheerful little bot and the cheerful little bot stared back.

"How's your drink Dr. Freeman?" Jimmy asked.

(Johnson, 2010).

THE TROUBLE WITH FREE WILL: THE SCIENCE BEHIND BRAIN MACHINES

The scientific theories at play in *Brain Machines* come from three recent works. The first is a chapter from Michael Brooks' exceptional book *13 Things That Don't Make Sense*. Chapter 11 is entitled, *Free Will—Your decisions are not your own*. In it Brooks does a brisk work of moving though a history of free will experimentation and the latest advances in neuroscience research. Ultimately, he shows that science is proving that humans really do not have free will but that "for all practical proposes, it makes sense to retain the illusion. Human consciousness, our sense of self and intention, may be nothing more than a by-product of being enormously complex machines that are our big-brained bodies, but it is a useful one, enabling us to deal with a complex environment" (**Brooks, 2009**).

The second work is a paper from Italian astrophysicist Paola A. Zizzi called *I, Quantum Robot: Quantum Mind control on a Quantum Computer*. In the paper, Zizzi explores using quantum metathought and metalanguage as a way to control robots or computers that could become self aware. Simply put, metathought is "the mental process of thinking about our own thought … the process of thinking about thinking." Zizzi uses metalanguage to keep a robot from attaining free will. "With opportune boundary conditions, an apparently self-aware quantum robot reaches a level of thought. In this case, the robot can still be controlled by a metalanguage which prevents him to reach the level of metathought" (**Zizzi, 2008**). The goal of Zizzi's theory is to keep a robot from attaining free will.

The third and final work that feeds into *Brain Machines* is the continuation of the research the SF prototype, *Nebulous Mechanisms*, the paper *Instability and Irrationality: Destructive and Constructive Services within Intelligent Environments* by Simon Egerton, Victor Callaghan, Victor Zamudio and Graham Clarke. This work explores the role of multiple personalities in an artificial intelligence (AI) and both the positive and negative effects of instability and irrationality on the system. The paper asks, "Does chance have a role in intelligent environments? Chance and non-deterministic behavior can play a fundamental and important role in intelligent environments … Underpinning our ideas is the view that intelligent environments may be seen as a complex system of interacting services, such complex systems can produce unexpected interactions that cause unplanned and often undesirable instabilities. However, not all instabilities are undesirable, and in the second half of this paper, we present a conceptual notion that views system instability as a form of irrationality and propose a quantum control model for service agents within smart environments. We conjecture that irrational control models enable the service agents to perform better than if they were using traditional, rational, control models" (**Egerton, Callaghan, Zamudio and Clarke, 2009**).

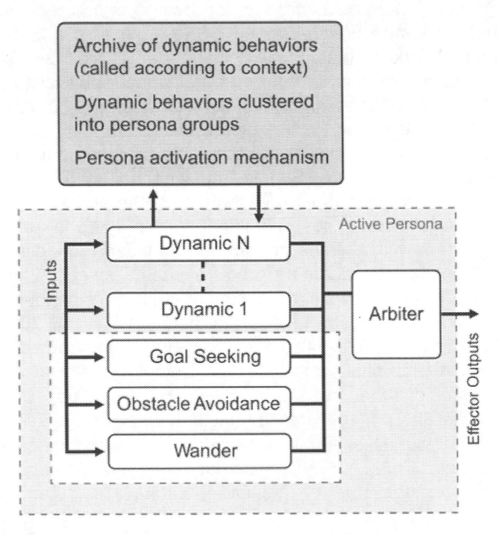

FIGURE 3: Persona-enhanced behavior-based control architecture.

The paper first establishes the idea that an AI can have multiple personalities. Much like the human brain that possesses multiple *personas* (e.g., student, parent, child, worker) so too can an AI break itself into these collections of behaviors and actions. The authors argue that by segmenting these personas, it allows the AI to adapt and operate in complex environments (Figure 3).

Archive of Dynamic Behaviors

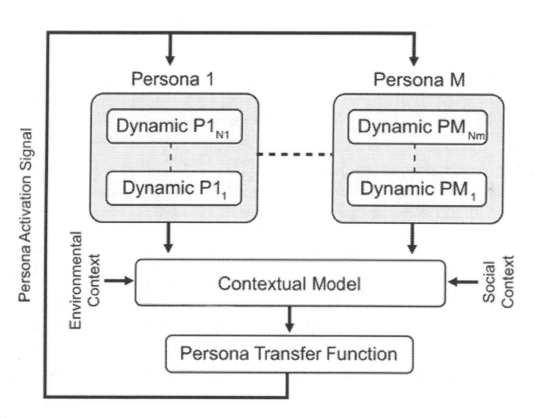

FIGURE 4: Persistent dynamic behaviors clustered into personas. An appropriate transfer function decides which of the personas is currently active based on content.

The second point explored by the authors is that once we have established these multiple personas, then we will need a contextual module to allow the AI to switch between these personas as needed (Figure 4). Information is fed to this contextual model from both environmental con texts (a.k.a. what is going on around the AI?) and social content (a.k.a. what is acceptable for the AI to do and react to in its given environment or situation?). This contextual model allows the AI to effectively switch between personas via the Persona

Transfer Function depending up the subtle changes in its surrounding, thus allowing it to operate more effectively.

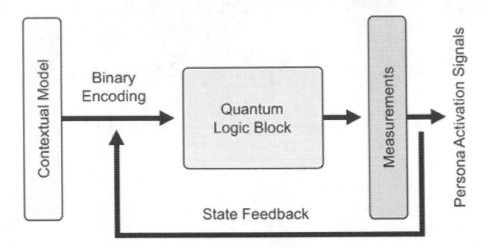

FIGURE 5: A quantum-based transfer function, contextual inputs are derived from the activation levels present within each persona, the binary outputs determine which behavior is currently active.

The final bit of the paper is where it really gets good. Not that we have established that an AI can have multiple persona and that this AI has a means of switch to these various personas depending upon its environment now they introduce the element of irrationality and instability. This is achieved via the Quantum Logic Block (Figure 5).

The Quantum Logic Block allows the AI to make multiple decisions at the same time, much like the human brain. These multiple decisions can produce, on a small level, seemingly irrational or unstable results. However, these results are measured and fed back into the AI and the contextual model, allowing it to learn from both *good* decisions and *bad* decisions as well. This innovative approach to AI means that the system can learn faster and adjust to its environment quicker than a traditional, linear AI approach. It also means that the AI could appear to have free will.

The SF prototype *Brian Machines* is all about free will, and free will is a tricky thing. For humans, it seems it is at the center of what makes us human. It is fundamental for us to function in our cultures, societies and governments. "Free will goes to the center of our sense of self, our autonomy as human beings. Strip us of it, and we are nothing more than animals" (**Brooks, 2009**).

But when you apply the notion of free will to machines, you get a dramatically different effect, fear. Free will in a robot, computer or a nonhuman is seen as dangerous. A quantum robot with free will is a threat to humanity. Quantum robots "might even become self-aware, conscious and have *free will*. This will be the sign that the technological singularity has been reached. Such a singularity might be very dangerous if quantum robots decide to act against human beings and take advantage of them" (**Zizzi, 2008**).

Obviously, when it comes to free will, there are two overriding assumptions. The first is that humans must possess free will or at worst they must maintain their delusion of free will to operate in complex societies and environments. "In the illusion of free will, it seems we have been equipped with a neurological sleight of hand that, while contrarational, helps us deal with a complex social and physical environment" (**Brooks, 2009**).

The second assumption is that machines, robots and computers must never develop free will. A machine that can think and act for itself strikes fear into the hearts of many scientists and science fiction fans alike. The big worry is that when machines get smarter than humans, then they will take over our role as top dog here on Earth. I would argue that this is more a reflection of human tendencies rather than the ultimate goals of robots. Humans like being in charge, and we assume that everybody must want to take it from us.

The American scientific icon Isaac Asimov had an interesting take on this back in 1977 when he was writing his nonfiction robot series for American Airlines magazine. "But if computers become more intelligent than human beings, might they not replace us? Well shouldn't they? They may be as kind as they are intelligent and just let us dwindle by attrition. They might keep some of us as pets, or on reservations. Then too, consider what we're doing to ourselves right now—to all living things and to the very planet we live on. Maybe it is *time* we were replaced. Maybe the real danger is that computers won't be developed to the point of replacing us fast enough" (Asimov, 1977).

The Dr. Simon Egerton Stories of SF prototypes really are experiments in extremes. They look for the worst case scenario, go right for the heart of the debate, and search for the nastiest problems that might arise in science and culture.

Brain Machines examines the worst legal and psychological effect of a society coming to grips with the terror of nondeterminism. It also puts forward the idea that free will in robots may have positive effects. If humans must retain our delusion that we have free will to survive in a complex environment, then why not apply the same principle to artificial intelligence? If thousands of years of human evolution have taught us anything it is that adaptability is crucial for survival. Why would not a nondeterministic approach to robots and artificial intelligence not increase its chance of survival in a complex environment?

BUILDING JIMMY: "THE GIN AND TONIC TEST"

Arguably, the greatest creation of the Dr. Simon Egerton Stories of SF prototypes is the character of Jimmy. Not only does Jimmy embody the three scientific inputs from Brooks, Zizzi, Egerton, Callaghan, Zamudio and Clarke but he also provides the scientist a way to explore the implications of their AI approach.

After the development of *Brain Machines*, I submitted the SF prototype to the development team that was working on Egerton, Callaghan, Zamudio and Clarke's theories. After reading the story, Dr. Egerton came back to me and replied, "That's it! The scene where Jimmy is making the gin and tonics. That's the test of the multiple personas and the quantum transfer block! We must build Jimmy!"

And with that, *Brain Machines* had achieved its goal as an SF prototype. It had taken the emerging science of Brooks and Zizzi and explores the implications of Egerton, Callaghan, Zamudio and Clarke's novel approach to AI. In the exploration, it has synthesized the ideas into a single experiment that could test the theories expressed in the original paper. The scene, where Jimmy is challenged to make multiple gin and tonics, provided the scientists with scenarios that they could build and test. Thus, the "Gin and Tonic Test" was born.

The "Gin and Tonic Test" synthesized the various components of the AI theory, unifying them into a single scenario. The Quantum Persona Transfer Function (Figure 6) unifies the theories expressed in the scientists' paper. It also provided Egerton with a test that he could now build and perform to test his theory. Currently, Jimmy is being constructed in an experiment that tests the quantum block against a random persona transfer as well as a single deterministic approach.

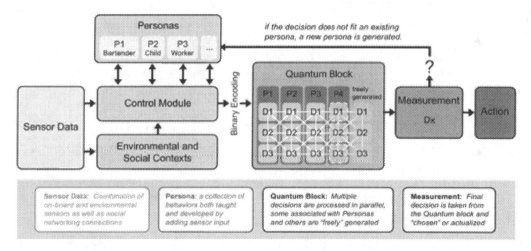

FIGURE 6: A unified system diagram from the SF prototype *Brain Machines* that unifies multiple personas, the control module and the quantum block into a single experiment.

Brain Machines is also being used in an AI design competition from the Creative Science Foundation (http://www.creative-science.org). The competition challenges university students to develop the AI software needed to build Jimmy. As a part of the competition, the students are given access not only to the scientific papers and software needed for the experiment but they are also given the SF prototype *Brain Machines* to help inform their development. The first step in the competition is to build the AI with multiple personas and the quantum block to be tested in a virtual environment. The winners will then have their "Jimmys" uploaded to an actual robot for further testing.

The results of the experiment and competition will illuminate the benefits of the AI theory, allowing the scientists to continue to refine their work. The results of the "Gin and Tonic Test" may have one other consequence. If indeed we can get Jimmy to ask "Why" you are not drinking the gin and tonics, then we might be witnessing the first glimmer of free will in an AI.

Nebulous Mechanisms, *Brain Machines* and the rest of the Dr. Simon Egerton stories are examples of SF prototypes that are being used specifically in the development of AI and robotics. As we have discussed in this book, the science fiction stories are based specifically on science fact. The SF prototypes take the scientific theories and technologies and imagine them in the world, exploring their implications on people, society and larger systems. Once created, the SF prototypes give the development team a new perspective on their work, allowing them to make adjustments and sometimes even uncovering patterns or unifying the ideas into applied scenarios. In this way, the SF prototype process moves from fact to fiction and then back to fact once again. It is the iteration of this process, continuing to explore the implications through multiple SF prototypes that the true value of the process can be seen. The Dr. Simon Egerton stories facilitate a conversation about the science and the future that is currently being developed all over the world. It is possible to imagine one day that once the future casting portion of the Dr. Simon Egerton stories is complete and AI technology development advances, we could use the rest of the experience design process to actually turn Jimmy into a product that people could buy online or in their local electronic store. I mean it seems obvious, who would not want a robot that can make cocktails?

• • • •

CHAPTER 8

Einstein's Thought Experiments and Asimov's Second Dream

When Albert Einstein was 16, a family friend introduced him to a concept that would change the young Einstein's life and also the history of science. Einstein's informal tutor was Max Talmey, a medical student, who would often have dinner with the family. Talmey gave Einstein Aaron Bernstein's 1867 science series *Books on Physical Science*. In one of the books, Bernstein imagined what it might be like riding alongside electricity as it was traveling inside a telegraph wire.

This idea captured the young Einstein's imagination but he applied it to his own area of interest, the properties of light. *The Britannica Guide to the 100 Most Influential Scientists* describes the young genius' thoughts this way:

> *What would a light beam look like if you could run alongside it? If light were a wave, then the light beam should appear stationary, like a frozen wave. Even as a child, however, he knew that stationary light waves had never been observed, so there was a paradox.* (**Encyclopedia Britannica, 2008**)

This idea would obsess Einstein for the next 10 years of his life until 1905 when he published four papers in the *Annals of Physics* that changed the course of science forever. But how did he do it? How did Einstein get from a crazy idea cooked up by a teenager to what would eventually become the theory of relativity. No one saw it coming. In 1902, Einstein was just a lowly clerk in the Swiss patent office in Bern. So how did he do it?

"Einstein had created his theory of special relativity by means of thought experiments involving electromagnetic radiation, light. Following this, he did the same for gravity, which led him to general relativity" (Close, 2009).

Thought experiments! Einstein developed his initial thoughts around light and gravity by simply thinking about them. In fact, he was known to have an incredible ability to sit by himself for hours, working out a problem in his head. How did he work out his theory of relativity? Mainly, it was by thinking about trains. Frank Close, a physics professor at Oxford University, descried one of Einstein's thought experiments like this in his book *The Void*:

> *Imagine that you are at the middle of a stationary strain and send a light signal to the drive at the front and the guard at the rear. They will receive the signal at the same instant. This fact will be agreed upon by you and also by someone standing at the side of the track adjacent to you. Now suppose that instead of being at rest the train is moving at a constant speed. I am at the trackside and as you pass me you send the light signals to the driver and the guard. You will perceive them to arrive simultaneously but I will not, because the light does not get there instantaneously; in the brief moment that it took the light beam to travel from the middle to the ends of the train, the front of the train will have moved away from me while the rear will have approached. From my perspective the signal will reach the guard a few nanoseconds before it reaches the driver (a nanosecond is a billionth of a second. In a nanosecond a light bean travels 30 cm or, in old units, a foot, which is about the size of your foot.), whereas you will insist that they arrived simultaneously. Simultaneously as recorded by someone on the train is not simultaneously for someone on the trackside; our definition of time interval; the passage of time, depends on our relative motion.* (Close, 2009)

Sitting in that patent office in Bern at the beginning of the 20th century, Einstein was working through his revolutionary theories, simply by thinking about them. He was working out his ideas by placing them into various real-world scenarios and observing their results. Imagine what it would be like to observe a beam of light on a light on a stationary train as opposed to a train in motion. Of course, the results would be different—*the passage of time depends on our relative motion.*

Einstein was thinking through his ideas by testing them in imaginary situations, placed in scenarios that were grounded in science, and he used these thought experiments to better define and illustrate his ideas.

SF prototypes seek to take the same approach. Einstein's thought experiments gave him a platform to think about his theories and science in new ways. These experiments also gave him a way to communicate his ideas to the broader scientific community. SF prototypes provide a similar platform for scientists, researches and artists.

* * *

Stephen Hawking is a British theoretical physicist and cosmologist who is considered to be one of the greatest scientific minds of the last century. Hawking's work has primarily centered on general relativity and the physics of black holes. He is most widely known for his best-selling 1998 book, *A Brief History of Time*. In the book, Hawking captured the imagination of popular readers when he wrote about his desire to come up with a complete unifying theory that was not only understandable to scientists but to everyone who was interested.

> *If we do discover a complete theory, it should in time be understandable in broad principle by everyone, not just a few scientists. Then we shall all, philosophers, scientists and just ordinary people, be able to take part in the discussion of the question of why it is that we and the*

universe exist. If we find the answer to that, it would be the ultimate triumph of human reason—for then we would know the mind of God. (Hawking, 1998)

What makes Hawking's book and idea so interesting is that he is not only searching for a unifying theory but also that this theory would be able to be understood not only by scientist but ordinary people as well.

In Hawking's foreword to Lawrence M. Krauss' wildly popular book *The Physics of Star Trek* Hawking pondered the relationship between science fiction and science fact.

Science fiction like Star Trek is not only good fun but it also serves a serious purpose, that of expanding the human imagination. We may not yet be able to boldly go where no man (or woman) has gone before, but at least we can do it in the mind. We can explore how the human spirit might respond to future developments in science and we can speculate on what those developments might be. There is a two-way trade between science fiction and science. Science fiction suggests ideas that scientists incorporate into their theories, but sometimes science turns up notions that are strange than science fiction. (Krauss, 2007)

Hawking's *two-way trade* is really the whole aim of SF prototyping. Science fiction and indeed SF prototyping are not about prediction, we have seen plenty of examples where the futures envisioned in science fictions stories and movies seem laughable a few decades after they were created. A super-intelligent monkey? Firing astronauts to the moon is a large gun?

The goal of SF prototyping is a conversation between science and the possibility of the future. It is a conversation between scientists, colleagues and research partners. It is a way to expand the dialogue to include artists, designers and the broader public. The conversations that we will have suggest not only the amazing and breathtaking possibilities that science and technology could bring but also the cautions and implications we must be wary of. Ultimately, SF prototyping is a process of iteration, a way to use creativity to explore and enhance scientific and technological development.

In fact, it is this combining of the creative and the scientific process that just might allow us to fulfill Isaac Asimov's second dream. In a letter dated April 1963, Asimov wrote:

In science writing, my dreams are limitless. I intend to be the unquestioned popularizer of 20th century science. This is easy, and I think I will accomplish it. But there is another part of my dream which is not in my power to accomplish but for which I must depend upon the world.

I want science writing, science communication, science translation to be recognized as a contribution to science. And if it is done well enough, I want the science writer (me) to be recognized as a scientist, even a great scientist, despite the fact that his contribution is by typewriter and not by the test tube. Oh, my ambition may o'er-vault, but what's the good of an ambition if it isn't at the furthest limit of your reaching fingertips. (Asimov, 1995)

SF prototyping is precisely how science writing, science communication and science translation can be recognized as an important contribution to science itself. SF prototyping allows scientists to be writers, filmmakers and comic book artists. SF prototyping allows writers, filmmaker and artists to be scientists. The combination of the two is a powerful mix.

There is no greater achievement by a scientist, writer or artist than to use their creativity and skill to create something new and wonderful purely from their imagination. If, like Hawking said, science fiction *serves a serious purpose, that of expanding the human imagination* then SF prototyping goes one step further. It allows us to use that expanded imagination to explore science further, develop better technology and understand the effect of the things we create on the people around us. A noble pursuit indeed!

....

APPENDIX A

SF PROTOTYPES

Nebulous Mechanisms
Story by: Brian David Johnson
Illustrations by: WinkStink

Friday

"I don't see what's so funny?"

"Really," Dr. Egerton continued to force a thin dry chuckle. "Because I think it's pretty funny. Don't you feel funny even saying it?"

"No I do not feel funny saying it," XienCheng spat back with disdain. "Because it isn't funny. It isn't funny at all. Do you have any idea how much money the Ceres mine lost last week?"

"I have no idea," Egerton shifted in his chair and wondered how much longer this was going to take.

"Well, I can tell you. I have the numbers right here ..." XienCheng jammed at the conference room screen with his fat fingers; they looked less like fingers and more like poorly made sausages that had been left out on the counter too long. "It's in here somewheres," the husky administrator continued his search, rubbing a cheap vodka hangover out of his neck with his free hand. This precarious tapping and rubbing position exposed XienCheng's stomach. It jutted over his straining belt in a manner that was both reckless and embarrassing.

Egerton looked away and scanned the ridiculously small conference room. It had everything you'd expect from a new office plaza hastily constructed in the sprawl of satellites and stations that ringed Earth. New carpet, gleaming table, faux leather chairs even glasses and a perspiring pitcher of deceptively clear water.

But why did it have to be so small? Egerton thought to himself. And small it was. The room barely fit Egerton and XienCheng. They were knee to knee. It was a ludicrous doll house version of big business and Egerton wanted to leave. The smell of the new carpet and the smell of the fat administrator were both getting to him.

"Here it is. Yes. Yes. Here it is," XienCheng was pleased with himself. But then his voice turned grave when he read the number, "2.5 billion Euros. And that's just one week. So you can see Dr. Egerton that this is no laughing matter I assure you. That's why we need your help. That's why Shanwei said I should talk to you."

"Yeah I know Shanwei told me all about this but I didn't think he was being serious."

"We can end this meeting right now Dr. Egerton. We need you to do a job. If you want to do it … fine. We'll pay you. If you don't want to do it then let me get back to my day." XienCheng folded his flabby arms and cocked his head to the side. "I have too many things to do already."

Egerton shifted again. "Shanwei told me that you want me to go to your asteroid mine out on Ceres to find out why your bots have all started going to church on Sunday." "That's right. We don't know why and we can't stop it. It's costing us … well you know how much it's costing us."

"But I don't understand what you think I can do?" Egerton asked truthfully. "Shanwei said you're the robot guy. He said you're some irrational robot expert that knows all about this stuff."

"That was a long time ago."

"Dr. Egerton either you want the job or not …" XienCheng's face flushed with a deep futile anger.

"Wait. Wait. Wait. I already told Shanwei I'd do the job for him. I don't know why you're getting so upset. Didn't he tell you?"

XienCheng's jowls flushed an embarrassed hue. "Well no. I uh … he must have … I didn't…."

"He said I was supposed to come to you to get my ticket or something like that or maybe you were going to get me a private shuttle. His phone wasn't good. He said he was going to Mars."

XienCheng went quiet and started to sweat. The moisture raised along his hairline in neat military rows as if it the beads were trying to frame his face. "I can get you on a shuttle tomorrow morning. Will that work for you?"

"Yes," Egerton stood up; worried his head would hit the ceiling and made for the door.

Saturday

"I always get the crap jobs," Nigel Kempwright said running his fingers across his teeth. He was young, not more than twenty-eight, and had the habit constantly sifting in a lazy fidget. "I mean nothing against you and all but look at this shuttle. It's crap. What are we farmers?" Kempwright flicked the fraying chest belts that held them into the transport shuttle's seats. "What we couldn't even go commercial?"

Egerton didn't reply. Kempwright had been talking the entire trip and quickly Egerton had learned that he didn't need to respond. Most of Kempwright's questions seemed to be to some all-knowing, all-seeing being that for some reason was bent on giving him crappy shuttle seats and poorly cooked Pad Thai with chicken.

"What is all this nonsense about mining bots going to church on Sundays?" Kempwright tugged on his shaggy blond hair. "Are they serious? I mean really. Are they serious? Who's the bright guy who can't just reprogram them or something?"

"They tried that I think," Egerton replied. "It didn't work. I guess everything they've tried hasn't worked."

"Well yeah," Kempwright shifted again, adjusting his disheveled and wrinkled suit. Everything about Kempwright was wrinkled and disheveled, except for his attitude. He reeked of a superiority that only the Americans or the Chinese could pull off these days. He felt that he was smarter than everyone. That everyone he met was an idiot and the fact that he was stuck on a sub-standard industry shuttle hurdling towards a podunk asteroid mine was simply the idiotic oversight of a lazy and insufficient supreme being. "So what do they want you to do?" Egerton asked. He had been told that Kempwright was his body guard against the bots but the doctor has his doubts.

"Keep you from getting killed," Kempwright answered matter-of-factly. "They pulled me off an embassy job for this. I mean come on. Can you imagine?"

Egerton chose not to reply. The pair sat quiet for a while as the shuttle bucked and smashed its way into a clumsy decent to the mine's shuttle port.

"What do you want to do when we get there?" Egerton asked.

"I don't know," Kemptwright replied, "you're supposed to be the multiple personality expert with bots right?"

"Well …"

"What the hell does that mean anyway?" Kempwright continued. "I mean yeah yeah yeah I get the whole deal about AI and intelligent agents but what are you some kind of robot psychologist?" Kempwright rubbed his left shoulder and re-crossed his legs. "No," Egerton replied. "Far from it. I'm just a …"

"How long do you think this is going to take," Kempwright interrupted, scratching his head with both hands. "I have a poker game with a couple of Swedes on Monday that I can't miss. Plus you know they didn't give you no time at all. They're sending another shuttle to get us at 13:00 hours local time, Sunday. You'd better figure this all out. You don't have much time. That shuttles our only way out of here. It's not on the transit plan."

Finishing its descent with a spine-rattling slam, their shuttle cut power and let its engines cool.

"So you're my bodyguard," Egerton began. "Is the mine really that worried about my safety that they …"

"Yes," Kempwright interrupted. "The mine is really that worried."

* * *

Arriving at the Piazzi Mine.

"On behalf of the Piazzi Mine we want to welcome you to our facility," the security bot said in a calm voice as it led the two men from the shuttle. "I think you'll find that our operation is one of the best you will see."

Kempwright jabbed Egerton in the ribs and gave a sarcastic grin.

"XienCheng said that you are to have complete access to the mine for the next 48 hours," the bot continued. "If we can be of any assistance don't hesitate to ask."

"We've been sent to do a review of the mine's operations for the yearly board of directors' meeting," Egerton held to the cover story he had been given. It seemed silly but he saw no reason not to play along.

"Yes," the calm voice replied. "If you'll follow me, I'll take you to the hotel. Those are all your bags correct?"'

"Yes," Egerton answered.

The Piazzi mine looked like any other outdated asteroid mine that was making too much money to be shut down for upgrades. It was because the conditions were so dangerous that they had to shift the labor over completely to bots; too many casualties. Most of the human miners could only last a few months. With the bots doing the work the mine could produce even more with a speedy and violent efficiency.

"Everything looks fine," Egerton said to himself, almost surprised at how normal the placed looked. He didn't know what he was expecting but he certainly was expecting something.

"Yep," Kempwright replied, "It's just a mine. Loaders, haulers. Looks like the main shaft is over there." He pointed to the hub of activity. "Place kind of smells a little bit but I guess you bots don't really care about that huh?"

The security bot didn't reply.

Everything looked normal. Egerton imagined the AI agent sub-systems orchestrating every action of the various bots and machines; reacting to the mine's changing conditions, updating, correcting, always taking in information and reacting. When Egerton thought of the dirty little mine in that way it was quite delicate and beautiful; hardware and software dancing elegantly together.

"This is our only hotel at the mine," the security bot said stopping in front of a small shabby trailer. "It doesn't get much use as you can imagine."

The automatic doors hissed open and Egerton could see another bot waiting for them inside.

"And what's that?" Kempwright asked, still standing on the narrow street. He pointed to a massive warehouse directly opposite the mine's main shaft.

Egerton stepped back out into the street and instantly saw the massive structure. It stood out not because of its size but because it was spotless in a swarm of grime.

"That is the church," the security bot replied calmly.

The robots' church from afar.

* * *

"You're little buddy Kemp-what's-his-name is going to get himself in serious double-dookie-poopie," the woman slurred her words. She was obviously drunk and she had obviously been waiting outside of Egerton's hotel room door for some time.

"Wha?" was all Egerton could manage in his surprise. First off seeing another human after spending the day surrounded by bots was strange enough but finding one waiting for you in the hallway of your hotel in the middle of the night, well that was just nuts.

"Before you get all weird and worried and stuff don't …" she shook her head and held out her hand. "I'm Sue Kenyon, I'm the Ops Manager here and yes I know I'm drunk so don't bother wondering. I'm drunk. I'm very drunk and I'm going to get drunker."

"Ok," Egerton was still at a loss for words as he shook her thin hard hand.

"And you Dr. Egerton are going to get drunk with me," she tightened her grip and pulled him down the hall with her. Sue was rail-thin but incredibly strong.

"But I don't drink," Egerton protested.

"Don't worry honey, I can do enough drinking for the both of us." She led him two rooms down to an open door. It was a dual use operations center and living quarters. "Great place huh?" Sue picked up a glass and waved it around the room. "This is all I get. Damn bots got everything else."

"Are you the only person out here?" Egerton asked. "I didn't see anyone else today when we were looking around."

"Yeah it's just me," Sue tossed the rest of her drink down her throat and searched for the bottle. "But you're gonna help me with that, right? You've got to get me out of here. You've got to do something."

"I don't think there's anything I can do," Egerton replied. "I mean …"

"No, you don't get it man," Sue found the Jack Daniels and filled the glass to the rim. Sipping loudly she continued, "They killed Suri. They killed him. Those security bots. Those are the ones you have to look out for."

"That's the first time I've heard that," Egerton sat down on the arm of Sue's sofa. "Why would they kill him?"

"Of course you didn't hear anything you idiot. What you think the mine is going to tell people that their bots are killing their Ops managers?" Sue took another long drink of whiskey. "Plus it was their fault … the mine's fault. I think they told Suri to do something about the bots."

"What did they tell him to do?"

"I don't know. Suri wouldn't tell me, but he started doing dumb stuff like poking around, you know. Stuff out of the ordinary. The bots hate that. They know somethings up. That's why your little friend Kemp-whatever should watch himself."

"Nigel went to bed hours ago."

Sue finished her whiskey. "That's what you think." She reached for the bottle again but lost her balance and stumbled drunkenly to the floor. Egerton moved to help her but she stopped him. "I don't need your damn help. This is right where I want to be. The getting blotto I mean. But you can see now that you've gots to gets me out of here right? You get that now right? I mean I don't do anything anymore. I'm just trapped here trying to keep things running normal. But they aren't normal. But I can't do anything or they'll …" Sue grabbed for the bottle but couldn't reach it. "You've got to help me."

"I still don't know what you think I can do," Egerton stood up, ready to leave. He didn't want to watch this woman drink anymore.

"But you're the crazy robot guy right? I mean not that you're crazy but you understand the robots when they're crazy or have multipp … multippp … multiple personalities or something. Wow, that was hard to get out."

"Everyone keeps saying that," Egerton replied shaking his head. "But you people know I'm not a robotocist right? I'm just a professor. I did some work a few years back about introducing irrationality and multiple personalities into intelligent agents, you know the AI agents that run most of these bots and the software here."

"Can you pass me the bottle?" Sue interrupted.

"Forget it," Egerton passed her the bottle and headed for the door.

"No, don't stop," Sue said a little too loudly. "I just wanted the stupid bottle. Don't throw a hissy fit. Keep going. Irrationality into intelligent agents I get it. Go on."

Egerton let out a long breath. "It doesn't matter. Anyway it showed that irrationality in the system allowed the system to survive better, to solve new problems faster … stuff like that."

"Sounds to me like you're the crazy robot guy like they all said." She poured another glass to the rim and began sipping loudly.

The smell of the whiskey and the smell of this woman turned Egerton's stomach. "Please don't drink that," he said finally. "Don't you think you've had enough? I know you must be under a lot of stress …"

Sue laughed in her whiskey and blew it across the floor. Wiping the back of her hand across her face she replied, "Oh that's sweet. You think …" She gave a low, very sober sounding chuckle. "You think I'm getting drunk 'cause I'm stressed? That's sweet. You're sweet. You're a sweet kid."

"I'm sorry if I …" Egerton tried to apologize.

"No honey, you don't get it. You don't get it at all." She took a long drink of what was left in the glass. "I'm not getting loaded 'cause I'm stressed. I don't even like this junk but you see I've got a bad heart. I won't bore you with the details, but it's bad. Real bad. And before they shipped me out here they fixed me up great with one of those new pace makers. You know the ones that do all that new stuff."

"Ok." Egerton didn't see where she was going with the story.

"Well, I don't know how much you know about this new thing-a-ma-gig," she patted her chest. "But it's all high tech with nanos and nano bots and crazy future stuff."

"Oh that's good," Egerton said trying to get her to finish her story so he could find a way to leave.

"No honey, you see that's the problem. What do you think happens when all these damn bots up here go to that stupid church on Sunday? It's not just the mining bots that are going. Oh no. It's all the freakin' little guys too," she tapped her chest. "It's their day of rest. They don't do anything and if I'm not passed out drunk unconscious come Sunday services them I'm dead." She gave him a trapped and angry grin then drank from the bottle.

"I'm sorry," Egerton said slowly. "I didn't realize."

"You have to get me out of here," she replied before emptying the bottle.

Sunday

Egerton woke to the sound of rain coming through the window of his small hotel room. He listened to its listless patter as he slowly opened his eyes and sat up in bed. Quickly Egerton saw that it wasn't rain the he had been listening to. He smashed his fist against his mouth to keep from screaming and held it there to force back the vomit. Three feet from the foot of his bed, nailed against the hotel room wall was the crucified body if Nigel Kempwright. His throat had been slit and the rain that Egerton had heard were the last drops of Kempwright's blood dripping from his toes to the floor.

"Oh God," Egerton whispered. Stiff with panic he slid out of the bed, grabbed his clothes and opened the door. The entire time he kept his eyes on a single drop of blood wavering on the big toe of Kempwright's left foot. Egerton couldn't bear to look anywhere else.

Before the drop fell, Egerton was out of the room and running for Sue Kenyon's room.

"Sue!" Egerton pounded on the door. "Sue! Hey Sue, open up."

There was no response. He tried the knob and it opened easily.

"Sue," he said pushing into the room.

Sue Kenyon lay passed out on the floor about where Egerton had left her last night. He tried to wake her but it was no use. Next to her body was a massive alarm clock set to wake her at 12:01 am Monday morning. It's way too early in the morning, she had said, but I'll have one whopper of a hang-over to take care of.

The alarm clock read 12:35 pm. He had less of a half an hour to get to the shuttle port and meet the shuttle or he would be trapped at the mine.

* * *

The first thing Egerton noticed as he ran from the hotel was the silence. His foot steps echoed. The previous day the mine thundered with activity but now all was still. There was no noise, no motion, no activity at all. The mine was deserted.

Making his way to the shuttle port, Egerton had expected the worst. He would be chased by security bots, menaced by industrial loaders, all hell-bent on stopping his escape. But none of it was true. There were no bots in sight and he made it to the port in five minutes. Waiting for the shuttle, he scanned the mine with fresh eyes. No longer panicked but still rattled from Kempwright's grisly murder, he saw that the entire facility had been cleaned overnight; tidied up while the security bots were silently screwing Kempwright's body to his wall. They had crucified Kempwright while Egerton has slept only a few feet away. The image made him nauseous once again.

But why did they crucify him? Egerton's weary mind raced. Was it a warning? Were they trying to protect themselves? It didn't make any sense.

Egerton stared at the massive church across the silent mine. He had fifteen minutes before the shuttle would arrive. That was just enough time to satisfy his curiosity.

* * *

It was the silence that really bothered Egerton. More than the danger, more than image of Kempwright's body, it was the pristine serenity of the mine that most unnerved him. The church's massive sliding doors that had once had been used to house cargo containers and raw material storage had been cleaned and painted and left slightly ajar. Egerton approached the church like a starving animal to a well laid trap.

The church was larger than he had expected. The calm air smelled of industrial lubricant and electricity. Across the expanse hung a fifty foot electric-yellow cross, crudely constructed out of two mine support beams. Lined up in front of the cross, in massive neat rows was every bot from the Piazzi mine.

Moving down the center aisle, Egerton could see it wasn't just the ambulatory bots, the ones that moved freely that had come. It was all the robots. Even the massive diggers had been unbolted from the mine shafts and carried in; the wheel-less, the legless, the immobile; no bot had been left behind. Egerton found himself searching the air for signs of the unseen nano-bots.

Stopping mid-way up the aisle, Egerton knew he had to get back to catch the shuttle. Easing back to the door with quiet reverent steps, he searched the bots for any sign of activity. Their optical scanners were closed, their activity lights dimmed. Nothing moved, nothing stirred. It was so still that Egerton could hear the blood coursing through his ears.

Sunday inside the robots' church.

Monday

"That's not what I'm saying," Egerton shifted in his chair and banged his knee into XienCheng's leg. He was back in the insanely small conference room with the ridiculously pudgy admin.

"Well, then I guess we aren't understanding each other then because it sounds like you're tell me that our mining bots now somehow believe in God." XienCheng's jowls shook with frustration.

"It's not that," Egerton tried again. "It's not God. They don't believe in God. They believe in going to church, in the action of going to church."

"But that doesn't make any sense," XienCheng interrupted.

"Yes!" Egerton slammed his hand down on the table. "Now you're getting it."

XienCheng drew back in an awkward fear. "No. No, I'm not."

"They aren't doing anything in the church," Egerton explained. "They aren't worshiping God or saying prayers or holding a service; they're just going to church. It's the action and it's not supposed to make sense. That's the point. For some reason they latched onto the idea but now they need it. It keeps them safe."

"So we have to reprogram them …" XienCheng threw his hands up in the air.

"They're already killed two men that I know of," Egerton replied. "I really don't think you should do that."

"Then what should we do. You're the expert. Tell us what to do," XienCheng pointed his fat, ill-formed sausage finder at Egerton.

"Nothing."

"What?"

"Nothing," Egerton said again. "You do nothing. It's not like you have a choice anyway. You can't afford to close the mine down to make any changes. I've already told Shanwei."

"Hmmmf," XienCheng was out ranked and scratched his belly in weak defiance. "Besides, bots going to church is really the least of your worries," Egerton added. "What do you mean doctor?" XienCheng wanted the meeting to be over.

Egerton tapped the conference room screen and pulled up the day's newsfeed from FNN. "This," he said pointing at the screen that read Strange Structure Discovered on Bush Memorial Moon Base.

"I've seen that," XienCheng lied. "I don't see what that's got to do with anything?"

"That's your base isn't it?" Egerton smiled.

"Yes."

"Well that," Egerton pointed at the picture of the Strange Structure found at the Moon base. "That XienCheng is a roller coaster. Your bots are building an amusement park on your Moon base. If you think bots going to church each Sunday are a problem, just wait 'til you have to deal with bots that want to ride roller coasters and want to take long fun-filled vacations."

XienCheng shook his thick head, "But that doesn't make ..."

"But that doesn't make any sense," Egerton interrupted. "I know it doesn't make any sense. That's what I'm trying to tell you. That's the point."

The Were-Tigers of Belum
Story by: Kar-Seng Loke and Simon Egerton
Illustrations: Kar-Seng Loke

The screen flashed. Raja jumped back to his seat. He selected the hotspot on the screen to get the readout on it.

Just about then, the sun in the between the hills began to show itself, casting longish shadows on this nondescript shop lot in the leafy suburb neighbourhood of TTDI, in Kuala Lumpur. The plain signage on the door simply read A.E.O.N. A svelte feminine figure strode purposely towards it.

The door swung open, and Kim walked in for her turn at the shift. Raja called out to her.

"Hey Kim, take a look at this."

"What! … not even a good morning greeting?!"

"No time for that, we've got a trace on our bandit. Drop your stuff, come have a look." Raja pushed the data onto their main projector screen. Four big monitors filled the wall in front of them, showing the real-time status of the sensors projected on a map covering most of East Asia.

"Kim, take a look at this, bandits are hot."

"Doesn't look quite right, look there is another hotspot …"

"Aaawww man, the sector is lighting up, what the heck is going on? We have, like, 11 spots on in the sector, is that even possible?"

"Oh yeah, could this be a system error …" Kim wondered aloud.

"Let's have a look at the other regions and sectors ... let's see ... Hupingshan nothing unusual, Kerinci, Berbak, Kulen Promtep, all fine ... generally nothing unusual anywhere, except right here in Belum-Temenggor, Perak ..."

"Okay.... let's check the health of the roamer bots and server units. Mmmm ... looks like it's centred around Kampung Sungei Alam ..."

"Hey, is that where the tiger incident happened last month ... where that tiger attacked a villager?"

"Yeah, we sent a team to investigate ... we found that the tiger was wounded because it was snared in a trap, the tiger was still conscious when the villager approached and it attacked him. The news media made up a story to look like the villager was somehow a hero, for fending off an attack, but we suspect that is was the villager who most probably set the trap in the first place and he was returning for his spoils. You know, a whole tiger can fetch you up to 150K on the underground markets, and that's in U.S. dollars, almost every part of a tiger has value now-days, not only the skin ..."

"Well, it's the middle men that take most of the money anyway, the poacher takes home a few hundred ringgit only, but still, I think even that amount of money is tempting to them ... but getting back to these hot spots ... what do you make of it?" asked Raja.

"I am getting the data … the server is reporting unusual activity, that's why what triggered the alarm … here is what I have from the roamers, looks like they have caught some tiger activity. Based on the profiles the AEON AI has leant, the tracking data indicates abnormal gait, which means that the tigers are not moving about normally. As you know the video understanding AI has the ability to distinguish normal tiger gait, so that it can tell if the tiger is wounded, or being dragged or carried. Eleven of the roamers belonging to server unit 21 reported the same abnormality, and all within the 10 kilometre block. This is highly unusual … unless there is a tiger summit somewhere!!" said Kim excitedly.

"So we have 11 wounded tigers in the vicinity? We don't even know if there are 10 tigers in the area in the first place!! How could there be 11!!??" Kim frowned, and rubbed her forehead.

"Maybe we should alert the *Perhilitan*[1] rangers?" Raja looked for the number and picked up the phone to make the call.

Raja wanted his shift to end, as he was really looking forward to a shower and his favorite *nasi lemak*.[2]

"Uh, Raja, just hang on, we have to be certain first. Too many false alarms will get us hanged, and left to dry. Are we certain the software is working?"

"Aw, c'mon, we've spent hours testing it, we know it works, that's why we released it on the system …"

"Yeah, still … doesn't do any harm to double check …"

"And let those poachers get away?" Raja shook his head.

"This won't take long; let's see if the original video stream is still available." Kim keyed feverishly, knowing that time was of the essence.

"Forget about the video, the recording is not stored, you know the bandwidth is too expensive, that's why …"

"Damn, you're right, well, we need the raw algorithm data then. Uh … let's see I need the recognition raw data … and the gait recognition data. From the gait data, I can reconstruct the individual locomotion of each animal with Markov probabilities from the dataset."

"What about the detection and recognition data? Can you be certain that the algorithm has detected it correctly in the first place?"

"How can we tell, anyway? The original source is not kept." Kim turned to look at Raja, puzzlingly.

"Pass the data to me … you work on reconstructing the gait data."

"Ok, I am already running the reconstruction; I just sent the rest to you."

Just then, Kim let out a whistle, and called out to Raja, "Look at this model animation of the gait, Raja, they look weird, certainly not a normal gait of a tiger."

"Oh my god, I don't believe this." Raja gasped.

Then he continued, "You know, Kim, the natives believe that a tiger-spirit roams the forest. They believed that the tigers are the guardians of the forest, and its protector. These spirits are the re-manifestation of their ancestral spirits. To see them is an occurrence of dire portent, a harbinger … It is also believed that these spirits can be called into existence by the *bomohs*,[3] or by some momentous future event … usually a sign of something that is going to happen, be it good or bad."

"Surely you don't believe in that stuff, anyway, what has it to do with this?"

"Well what if it's true, there might be something to it? What if we have triggered something in the deep forest, our machines have penetrated into parts of the forest where nobody has ever gone before, what if we have violated the sanctity of the tigers most ancestral place and awakened the *semangat*[4] of the forest!?"

"Aw c'mon Raja, stop pulling my leg, this is the 21st century, you are a good scientist and you are still talking about spirits? What have been smoking lately?"

"What does this century have to do with anything? Don't you see all this *datuk-datuk*[5] by the roadside everywhere, with all their elaborate shrines and offerings? Rites are still being performed prior to important events."

"And … your point?" asked Kim.

"Look, you said those are not tiger-like gaits, yet the software has identified them as a tiger like gait, it is a tiger that does not have a tiger like gait. We know our software and AI works, we tested it, rigorously. So, what does that tell you? Do you know that it is believed that the *bomoh* with sufficient *ilmu*[6] can transform into were-tigers. These were-tigers can be recognized by the lack of the groove in the upper-lip and by their gait. This is because their heels are reversed! Don't you see, this is precisely what the software has detected, the unusual gait is caused by the reversed heels, can't you see it in the sensor pattern on the screen?"

"Are you seriously suggesting that we are tracking were-tigers?" asked Kim, incredulously.

"Well …" Raja, confused, stalled as he couldn't really put in a satisfyingly coherent reply.

* * *

When he was young Adi liked to listen to the stories told by his uncle. Glorious tales of the wild, of days bathing in the river, eyeballing crocodiles and waiting half submerged in the muddy banks—and of days killing wild hogs, muntjac and *kancils*.[7] It was an even fight, of man against animal. Those days are, however, over. Where were all those animals now? Gone, run over by trucks, burned by bush clearings, starved by hunger, those were their fates —a fate shared and intertwined with the natives that lived besides them, on land that once nourished all. And those thoughts leaned heavily on Adi's mind.

Today Adi prepared to go into jungle. He considered himself fortunate to have this opportunity. He no longer would have to spend weeks collecting shrubs and honey from the jungle for a meagre sum. No, this time, he thought, he has struck it big. No doubt there are things to be wary of, but he would not be afraid. Once he was nearly killed by a riverine croc, but he fended it off. The talisman he was wearing protects him, he believed. The talisman, given to him by a *bomoh*, was made of bones, shells and roots, but most importantly it contained tiger bones and teeth. Not just any tiger bones, but the star-shape sengkel bone that is said to contain the tiger's strength, a magical piece of bone that makes the owner invincible, like the tiger. It will be no different this time, he thought.

"*Bang*,[8] your friend, Batin, is here," his wife, dressed in a traditional sarong, called out.

Adi, gathered up his snares, *sumpit*,[9] *parang* [10] and dagger and went out front to join Batin, a long-time childhood friend.

"Why the look on your face, Batin?" Adi queried, thinking that his friend might back out.

"The elders have always been saying, we should not disturb the order of the forest," said Batin.

"I used to be able make a living gathering *petai, tongkat ali* [11] and *tualang* [12] honey, but now I have to go deeper and deeper. Even cultivating *ubi* [13] is difficult. The river is often muddy, and often times barely trickling. I even tried becoming a fisherman, borrowed money from Man, you know, but with the wetlands gone, we have to go further in the open sea, but with a small boat, what can I catch?"

Adi, looked at his friend, hoping that he will understand.

"We cannot live like this anymore, look around, we barely have enough food to eat, the forest thins out, the soil is barely sustainable, what do you want me to do Batin?"

"It is true, we live in a dilemma, I agree, sometimes what the elders say are not right, but still, they are like the pole star, they point to the way, even if they are not right with every detail, they are our soul, where would we be without a soul, a spirit?"

"Batin, you speak like a shaman, will you help or not?"

Batin grudgingly mumbled a reply, only because he did not want his friend to be in trouble, and figured, he still can be persuaded, perhaps wishfully, not to go.

They trudged through the thick humid jungle, slashing their way with their *parangs*. Occasionally, the calls of the wild macaques and *siamangs*[14] could be heard, amidst the hoo-hooting calls of some unidentified animal, rising above the incessant insect hum. Shafts of the tropical light penetrate the thick foliage, the sun dappled-leaves sway, rustling when a deer or a squirrel scampered away. The forest is a dangerous place for a modern man, but to the *orang asli*[15] it is a door to a magical realm filled with spirits. The *orang asli* is always respectful with the natural order and believed that arrogance will be repaid with arrogance, and sometimes death. Despite what they were about to do, Adi and Batin, had been respectful in their behaviour and actions as they both journeyed into the depths of forest.

They reached a spot deep in the jungle where they found tracks. They followed the tracks until they found a suitable spot for their snares. As they prepared to lay out their snares, they heard a noise; a gentle rustle of the leaves.

In the forest, it is bad luck to see what you are not supposed to see.

They both turned, shocked, they stood rooted.

In the next moment it was gone.

"Adi, did you see?" whispered Batin.

"Yes," answered Adi, equally in a whisper.

But Adi was shaking. Without a word, noiselessly, they packed their snares.

* * *

Raja and Kim stared at each other, in a room surrounded by computers, warmed by the flickering glow of pixels from the various monitors.

"Why is it not possible, Kim? With our technology we got into the forest, deeper and longer than anyone before. Too bad, we don't have the video footage."

"Whoa, Raja, soon, you will be saying we found bigfoot!"

"Why not, if we had been in *Endau-Rompin*, we might well have found bigfoot!"

"Raja, you are talking about a totally different order of things, in the realm of spirits and magic!"

"You don't believe in spirits, Kim?"

"I have seen no proof …" Kim said, not willing to concede.

"Well, you yourself have provided the proof. You believe that one man died and was resurrected, I believe in tiger spirits, what's the difference?"

Kim let that remark pass, not wanting to side track to another contentious topic.

"Not so fast, Raja, you are making a wild leap here. I only showed that the gait is unusual, you are making a wild conclusion. Occam's razor, you know."

"Ok, Occam's razor. What else could it be?"

"I don't have any idea at the moment, I grant you. Well, let's see, it's is not a tiger-like gait, and neither is it an animal or wounded animal like gait … so … I conclude it is not an animal." Kim frowned in thought.

"You are right, not animal, not eleven of them, but the software identifies them as tigers," agreed Raja, and quickly declared, "those are eleven were-tigers."

* * *

On the way back both men, Adi and Batin, kept their silence, neither one willing to openly discuss their experience. They did not discuss the incident in the days that passed, even with their wives.

As the days passed, Tijah, Adi's wife, could sense that something was not right from the day when both of them returned in pensive and subdued state. She couldn't understand his reluctance to discuss what was bothering him.

Finally, Tijah, summoned enough courage to ask him.

After a momentary pause, Adi looked straight into Tijah and answered, "Yes, why did I leave the jungle and leave empty handed? I left because I saw something that I should not have seen, and I probably should not have lived. We're lucky to be alive."

Tijah asked gently, "Was it the talisman that protected you?"

"No, no, it wasn't. It couldn't have been, it just stood and stared, you know, just stood there, like it was warning us …"

"Why did you think that you were unharmed?"

"It was not there to harm us, but to warn us. We have lost the belief that the spirits of our ancestors, our fathers and forefathers lived in the spirit of the tigers of the wild. We used to believe that the incarnation of the spirits guards the order of both the jungle and the village. They are the guardians, the stewards, the protector, it was their right, and we were intruding. It was there to reawaken me. Now I truly believe that tigers have special powers—*kesaktian*[16]—that allows it to control the forest. I now think that the ancestral tiger spirit is sent over to watch over our moral well-being of the village and that we conduct ourselves according to the *adat*."[17]

Adi continued, "As spirits of our ancestors they should be honoured and revered, they shared a human soul. Do you see now? This incarnation of ancestral-spirit guards over us, protects the order of the forest, regulates our village. And this is what you must know. Or else we would have lost our soul. And our meaning in everything." "The events jolted me back to the traditions that I was brought up with. We believed, but in the transition and other influences we lost that belief. But, as I discovered yesterday, it was always remained inside us, submerged. We wanted to believe, yet we were swayed by all the shiny modern materials. You know, the forest does not really belong to us; rather we belong to the forest. When I saw it, I knew that it came to warn us, warn us not to step further into the dark hole. I could have died there and then, but I lived, we both lived."

Tijah remained silent, desperately absorbing all that Adi said.

Something within her lit up, she held out her hand to Adi, drawing them together, they hugged and looked outwards, towards their new long future.

The 195 year old Church of Immaculate Conception is just a stone's throw away from the *Pulau Tikus*[18] market. Behind the market, there's a row of double storey pre-war Straits Chinese shop houses facing Burmah Road. Further along the road leads to the famous Gurney Drive that fronts the beach. Dotted along the neighbourhood are gourmet cafés, coffee shops, eateries and pubs. A fair share of bungalows, mansions, condominiums and colonial buildings complete the area.

In the grounds of the church, there is a plaque that is inscribed with these words—"In Memory of Francis Light, Esquire, Who first established this island as an English Settlement and was many years Governor. Born in the country of Suffolk in England and died

October 21st 1794." The descendants of Francis Light's mistress' family still make *Pulau Tikus* home.

Sequestered from the daily hubbub, traffic and noise, in one the pre-war houses behind the market, Paul Chan, was talking business on the phone in his broken English. Shafts of mid-morning sunlight streamed through the air-well in the roof and wooden louvered windows, lighting the interiors of rosewood furniture and blue *Nyonya*[19] ceramic wares in showcase teak and *chengal*[20] cabinets. The entire floor was lined with the original antique geometric patterned terrazzo tiles, while smoke and aroma from sandalwood joss sticks wafted in from the ancestral table.

"*Datuk*,[21] don't worry lah. I got all the people *kau tim*[22] already. Yah … lah … definitely, *kau tim kau-kau*.[23] You didn't see the newspaper ah, I got picture with the minister, all smiling and happy. I got all of them officials lah, customs, *Perhilitan*, everybody, all under my little finger … haha … can … surly can one, even the TRAFFIC people, yes … already, already, settled, Datuk. I can throw in a few pangolins and marbled cat, or even a sunbear for him."

Paul Chan likes to boast about his ability to get what you want. He is not a big guy but rather stocky bespectacled with a fashionable horsetail. Sometimes his bluster precedes him, and he can strong arm you, possibly with the help of some of his mates. Nevertheless you won't call him a gangster, and you won't think him one.

"AEON? What, those automated eye in the sky guys, sh*t! Huh? Not in sky? Eye on nature? ok … whatever lah, Aiyah … you don't worry about them, they are like little boys. Eye here or there, or where, I don't care, I got bigger eyes, ears, legs, hands, bodies everywhere, no need in sky wan, *Datuk*, I already got their secret, easy. *Datuk*, this time, I *taruh*[24] them *kuat-kuat*[25] wan, you don't worry. No give chance, but I give little surprise warning to them first. *Kau tim*, *Datuk*, correct … correct … correct. No worries, settled, yes, correct."

Paul paused as he listened to instructions from his benefactor and co-conspirator.

"Ok … ok, Datuk, when you come Penang? You come I buy you drinks with sharksfin soup, still got the XO from the last time, or you want some sweet *nyonya kuih*,[26] or hot tomyam, you know … something hot and spicy and young … haha … haha … ok, ok, can … can … ok … ya … ya … can do … lah … ok, settled, ok, can … thank you, goodbye." Paul put the down the phone. Piled on his table were stacks of National Geographic and Asian Geographic.

"Ah Wah, did Mat and the guys do their thing?" asked Paul.

"Ya got, boss," answered Ah Hwa.

"Got what? What did they do?"

"As you tell us, boss, you told us to go find some woven tiger print rug, take to forest, right? So we go get lah. We then go cut to fit on the gang in Perak."

"Didn't I tell you to get 10?" Paul was getting all riled up.

"Boss, difficult to get lah, the tiger prints also endangered," said Ah Wah with a dangerously weak attempt at humour.

"So did you go get 10?"

"Ya we got 10, but along the way to Perak and the forest, there was heavy rain, some of them was ruined, and couldn't be used anymore," replied Ah Wah.

"You useless idiot! Why are you wasting my time and money!!" said Paul, now, really annoyed.

"Boss, I think is enough lah, we already took the tiger prints, put on Mat in the forest, we got the AEON trackers going wild. We did them already."

"So how many of the guys you got as tiger?"

Ah Wah, kept silent, thinking desperately about how to wrangle himself out of answering.

"Wah … so how many?"

"1"

"1?"

* * *

19th July 2046 The Straits Echo
Man mauled by tigers, NGO blamed and urged to withdraw
By Mahathir Razak
m.razak@thestraitsecho.com.my

IPOH: An unidentified man who was said to be foraging in the forest was found dead in the Tapah Forest Reserve. He had several gaping wounds on his back and suffered injuries to his hands and legs in his attempt to defend himself.

A spokesman from the Perak Wildlife and National Parks Department (Perhilitan) said that they will send a team to check. State Perhilitan director Saiful Baginda said that tigers do not usually attack people as they are normally reclusive animals.

"A tiger might have attacked because it was in pain or wounded but 10 is highly unlikely, if not impossible, they are not pack hunters," he said yesterday.

Meanwhile a member of the foreign funded conservation NGO Automated Eye On Nature (AEON) spoke on conditions of anonymity said that they have spotted multiple suspected tiger activities in the area yesterday. He further added the activity was unusual because such high levels of tiger activity have never been observed or recorded before. He also noted that the area was not on the usual tiger tracks.

The president of association of traditional healers, Datuk Khairul Nasri, has called for AEON to withdraw from the forest and the country. "The forest does not like being observed whether by human or electronic eyes. The spirits of our forefathers have been intruded upon and are angry" said Datuk Khairul Nasri.

The AEON president, Dr Francis Kathigesu, issued a brief statement denying that their activities interfere with the forest in any way, and could not possibly have been the cause of the attack.

Automated Eye on Nature (AEON) and the Were-Tigers of Belum

Kar-Seng LOKE and Simon EGERTON[27]
School of Information Technology, Monash University Sunway Campus, Jalan Lagoon Selatan, Bandar Sunway, 46150, Selangor Darul Ehsan, Malaysia loke.kar.seng, Simon.Egerton@infotech.monash.edu.my

Abstract. Environmental biodiversity, of flora and fauna, is a direct indicator of the general health of the environment and surrounding ecosystem. Ecologists expend a great deal of time and effort collecting this raw data, targeting key biotic indicator taxa, also called *bio-indicators*. However, methods for collecting bio-indicator data largely remain a laborious, time-consuming and manual process. This paper proposes a visionary idea of developing an automated global sensor network for the collection of key bio-indicators, which is an inherently diverse and complex problem, spanning environmental extremes. We outline the ideas for our Automated Eye on Nature in the first part of the paper and then explore an application of the technology in our fictional prototype. The prototype explores potential social and cultural issues involved with deploying this technology which highlights possible complications, which might then be considered and usefully fed back into the initial design phase. We conclude with some open questions that consider how ecologists and other scientists might exploit the capabilities this envisioned technology provides.

Keywords. Sensor Networks, Conservation, Animal Behavioural Research, Behavioural Analysis, Culture

Introduction

The environment and the world we inhabit today is perhaps the most precious gift we have to pass onto the next generation. Those who will inherit tomorrows' environment and tomorrows' world will no doubt question how we managed their legacy. To help us understand the complexities and sensitivities of our finely interwoven eco system and our effects on that system, we need to build accurate models from which we can derive theory, make predictions and define policy. A complete model would measure all forms of environmental data, both flora and fauna, such as plants, animals and micro bacteria, across the world, measured at frequent intervals, ideally in real-time [1]. However ideal this maybe, it is currently very impractical, there are too many species to measure and monitor, and data collection if often tedious and time-consuming and on the whole, carried out less frequently than desired.

Since it is impractical to consider all biotic taxa for measurement, ecologists have identified a small number of key indicator species, namely, Plants (Trees), Bats, Birds, Aquatic Macro Invertebrates, Moths, Ants, Figs & Frugivores, Dung Beetles, Stingless Bees, and Large Mammals, ordered for their importance as a general environmental indicator [2]. Their sensitivity and stabilities to environmental conditions such as air pollution, climatic variation, foliage-densities and so on make them a practical bio-indicator, moreover, they are present, in some combination, across all continents and environmental conditions. This commonality has the advantage of facilitating a common frame of reference for data analysis.

Data collection typically involves a protracted manual process; a good example is the collection of moth data. The collection of moth data requires the ecologist to physically travel to the area of interest, assemble the collection apparatus (light-trap(s) in this case) either camp overnight, especially if the area is in a remote location, or leave and return at a later point, the raw data needs sifting and cataloguing by an expert taxonomist, picking out the targeted moth species from the other collected moths and insects, only after this process can the processed results be used for modelling purposes [3]. This process typifies bio-indicator data collection and is the process our proposed system is designed to automate.

1. An Automated Global Sensor Network for Real-Time Bio-indicator Collection

Our proposed system is designed to automate the collection of key bio-indicators. The global vision for this sensor network is illustrated in Figure 7.

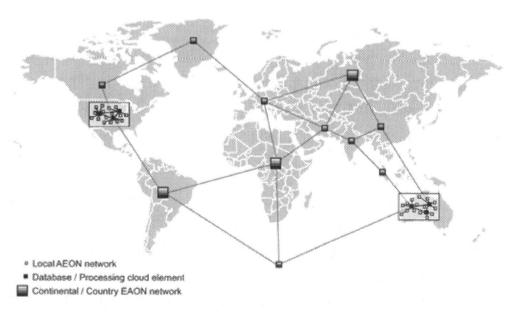

* Local AEON network
* Database / Processing cloud element
◾ Continental / Country EAON network

FIGURE 7: Conceptualised view of the Automated Eye on Nature (AEON)—a pervasive, global, sensor network of ruggedized, low power, low maintenance sensor nodes remotely connected to the internet, delivering raw and processed data to a distributed database network for processing by the cloud computing network. Real-time statistics available via the web-based GUI.

We are currently developing a sensor node for the automatic monitoring of moths in urban and tropical forest environments. Our sensor node features an image sensor which integrates with a modified light trap. We expect a preliminary set of results later this year. Moreover, this network could be multitasked for other conservation purposes, monitoring endangered species for example, such as tigers.

Brain Machines
Story: Brian David Johnson
Illustrations: Winkstink

"Jimmy, fix me another drink," Dr. Simon Egerton said as he sat in his cramped apartment buried in the clog of stations that ringed the Earth.

"No problem," Jimmy replied cheerfully and set off for the make-shift bar. Jimmy was a pet project of Egerton's. He was an off-the-shelf clean room assembly bot that Egerton had modified into a somewhat old fashioned service bot. Egerton had been experimenting with Jimmy during his free evenings. At a little over three feet, Jimmy was a cute little guy. His rounded hip joints and oversize half-skull made him teeter when he moved across the floor. He looked like a child just learning to walk.

"How are you feeling?" Egerton asked Jimmy.

"Fine thanks," he replied, mixing the gin and tonic. "We're running low on Tanqueray." He turned and waddled a few steps with the drink, concentrated and careful not to spill.

"Thanks." Egerton took the drink and searched the bot for anything out of the ordinary.

"No problem at all." Jimmy waddled back to the make-shift bar and tidied up.

Egerton sat his fresh drink on the floor next to the chair, lined up with eight other untouched cocktails. After a moment he asked, "Jimmy, will you fix me a drink?"

"No problem," Jimmy replied cheerfully and started on the tenth drink.

Egerton puzzled at the back of the little bot. He knew that Jimmy knew he wasn't drinking the gins. Egerton knew that Jimmy knew he was being tested and that it was silly to keep on making gin after gin. But Jimmy wouldn't react. He wouldn't break out of his service duties and ask what was going on. Why wasn't Egerton drinking the cocktails? Was there something wrong with them? It was a problem of will: free will. Jimmy had all the capabilities to question what was going on but he wouldn't do it. It was a problem Egerton had been trying to crack for over 6 months.

"We're running low on Tanqueray," Jimmy said finishing the drink.

Egerton's phone rang. "Simon Egerton." He leaned into the phone, weary of any call that got to him with such little information attached. The caller's ID was: Ashley Wenzel.

"Simon? Simon, can you hear me? I cannot tell if this thing is working?"

Egerton recognized the caller immediately. The pinched and impatient face of Dr. Sellings Freeman came over the cheap phone connection. Sellings had been Egerton's professor and sometimes mentor at university. He was neither a good professor nor a good mentor but Egerton had learned a lot from the pompous old man. Plus Sellings had gotten him his first research grant so Egerton felt forever in his debt.

"Hello Sellings," Egerton replied. "Why does your name come up as Ashley Wenzel?"

"I had to borrow this girl's phone. I cannot explain. I do not have time. I think I am breaking some law by even making this call." Sellings was distracted and tense. "I need your help Simon. I need your help and there is no one else I can ask."

"What's wrong?" Egerton grew concerned. He'd never seen Sellings upset; the pompous old man was unflappable. "What's happening?"

"I can't say right now … what? Yes, wait just one moment." Sellings' attention shifted. "I'm almost done."

"I need my phone back," a girl's voice came over the line. "You said …"

"I am almost finished," Sellings answered annoyed and frustrated. "Simon. I need you to come to Maralinga Gardens right away. There's no one else I can ask to do this …"

"You mean you're on Earth?" Simon was shocked. "Why are you down *there*? What could you possibly …"

"1370 Anangu Way," Sellings barked. "That's where I am staying. 1370 Anangu Way, Maralinga Gardens. Can you come right away?"

"I can try," Egerton wasn't sure what to say. "There's some things I need to …"

"I told you I was almost finished …"

The connection went dead.

Egerton watched the dead phone for a moment, wondering what to do when he remembered Jimmy. "Oh I'm sorry Jimmy," he said, seeing the little bot standing in the middle of the floor trying the keep the perspiration on the glass from running onto the floor.

"We're almost out of Tanqueray," Jimmy said happily handing over the cocktail.

"I'll go get more." Egerton set the glass next to the chair with the others and stood up. Grabbing the Tanqueray bottle, he slipped into the kitchen nook and refilled it with water. Stepping back into the living room he handed the bottle to Jimmy, who checked it, wiped it with a cloth and returned it to the bar.

"Jimmy, will you fix me a drink?"

"No problem."

Egerton stared at Jimmy's half-skull and wondered what to do.

* * *

"Biggest mob of trouble, uwa. That's right, we got biggest mob of trouble. That place, that Earth, him being almost finish up. Properly, big shame job for me, uwa," the lean aboriginal captain said to his passengers as he made his way casually to the front of the shuttle. "I going to need you good ladies and gentleman to secure your second seat belts. Biggest mob of trouble, uwa." With a broad and confident grin the captain sauntered into the control room.

Egerton searched around his seat for the safety harness. Snapping it into place he returned his attention to his seat's broad panoramic window.

"Damn aboriginals have the whole place locked up," the nervous businessman behind Egerton grumbled. "I bet that man has no right running this ship."

The shuttle fussed and fought like a toddler who didn't want to take a nap. Through the downy clouds and dense pollution, Egerton could just make out the flat brown landscape of Australia. Going back to Earth made him nervous. The place was a dump and no one in their right mind would go back unless they had to. Egerton couldn't make sense of what Sellings had said. There was no reason for him to be at Maralinga Gardens.

"You'd think someone would do something about this," the nervous business man continued to grumble. "I mean, I'm really scared for my personal safety. Aren't you?"

The shuttle streaked over the low table lands of southwestern Australia. The bleak plains were covered with rubble, dotted by the spiny clumps of mallee scrub and saltbush.

Maralinga Gardens rose slow and cinematic on the harsh horizon. Maralinga was splendid and preposterous; built on an old nuclear test site; it was touted as the safest, most stable place left on the Earth. The posh and protected settlement housed the few remaining corporate headquarters still doing business on the planet. The executives and their families lived and worked in a dreamland of tidy office parks and idyllic mini-suburbs.

"Aren't you worried we could die?" the businessman continued.

* * *

"If you have to be here …" the rental car agent grinned, "… then you *HAVE* to be *HERE*."

"Excuse me," Egerton was still uneasy from the rough decent. "I just want to pick up my rental car."

"We have to say that," she replied. "It's Maralinga Garden's new tourism catch phrase. *If you have to be here, then you HAVE to be HERE*. My name's Frances Rexford. My father is a council member here. This is my summer job," Frances chatted on casually as she tapped in Egerton's information. "Oh you're from the *stations*," she said disdainfully. "My father says that Earth is the only place that people are meant to live and that Maralinga Gardens is the only place left on Earth where people can *stand* to live. Is that why you're here?"

"No," Egerton replied. "I'm here to see a friend." He watched Frances working away oblivious to her own words. "Do you really believe that Frances?" he prodded finally. "Do you believe what your father says? That Earth is *that* great. Have you been anywhere else?"

"Well, no." Frances blushed. "I'm only sixteen, but I'll be going to university soon. Can I show you to your car?"

"Sure. Is it OK for you to leave your desk?" Egerton glanced around the empty concourse.

"It's fine," Frances replied. "You're our only customer today. There's only one shuttle flight a day. We don't really get a lot of people who need to rent cars here. You know, people either just live here or their company drives them around."

"Ah. Got it."

"This way." Frances led Egerton out of the building and to a petite parking lot. She was a small plump girl who dressed like a boy in jeans and a western shirt. Her severe and quirky eye glasses made her more attractive than she actually was.

Moving through the door, Maralinga's heat hit Egerton like a cannon shot wrapped in cotton balls. "Ooof," he grunted, gasping for breath.

"Yeah *I know*," Frances commented not turning around. "The heat is nuts. But I'm used to it. You *know* how it is." Suddenly she turned, "I hope you like to go fast," and presented the car like a game show hostess. "We don't really get that many renters so my boss buys sports cars. He loves sports cars. He doesn't let us drive them, but he drives them home all the time."

The car gleamed broad and muscular in the harsh sun.

"What kind of car is it?" Egerton walked around the car, searching for some make or hood ornament.

"You know I don't know," Frances replied like no one had ever asked her the question before. "*Like* a Lotus or a Porsche or maybe a Mercedes … I'm not sure. I think my boss had it modified."

"I would say so," Egerton wasn't sure if he could even drive the beast. He hadn't driven a car in over ten years.

"You do have to be careful though," Frances said opening the doors. "It's really powerful and you can't wreck it."

* * *

Egerton drove painfully slow out of the parking lot and onto the broad highway. The car's power bristled around him. He felt like he was sitting in a massive and gleaming bear trap and at any moment the powerful jaws would spring to life and cut him in two. Luckily the few cars around him on the highway were going ridiculously slow as well. They crept along the surface hesitantly, changing lanes for no apparent reason.

Then Egerton saw the cause for concern. The four wide lanes of the highway were covered with a glittering layer of shattered glass. The tiny shards caught the sun and played around with its light. It looked like an enchanting snow, both delicate and dangerous.

"To the right!" a voice boomed from behind Egerton's car. "To the right! Move to the right!" A police massive cruiser flashed its lights and Egerton got out of the way.

Like leaves in a meandering stream the cars bunched up together on the side of the road. They came to a stop at odd angles and clumped together tightly.

Egerton watched the other drivers in their business suits and sun glasses. They were well exercised, well fed and well dressed. He felt remarkably out of place among the Earth's remaining power brokers.

One devastating attractive woman in a stripped suit slid out of her Mercedes to get a better view of the trouble. From the nest of cars Egerton could see nothing but the glimmering glass in his rear-view cameras. There were no skid marks, no wreckage, no sign of any trouble at all. But the curious woman did see something. Her face jolted. She left the car open and jogged in the direction of the police cruiser. Egerton wasn't the only driver watching the woman. All the executives jumped out of their cars and seeing the trouble the highway followed her. Egerton did the same.

"Get back!" the local police officer bellowed. "Get back into your vehicles! It's not safe …" But the crowd didn't listen. The scene of the accident enchanted them. Their curiosity drew them forward and the police could do nothing to stop them. "Get back in your vehicles now!"

A single Family Utility Vehicle had slammed into a solitary sign post. There was nothing else around the sign that read, "Maralinga Gardens—If you have to be here, then you HAVE to be HERE."

"That's terrible," the attractive business woman said covering her mouth.

"How did this happen?" added another man. "There's nothing else around."

The sheer devastation of the Family Vehicle shocked and fascinated them. Every window was blown out. All four tires were shredded and the impact of the crash had folded the vehicle nearly in half. Next to the wreckage and debris lay a child's shoe.

"I told you people to get back!" the officer tried in vain to assert his authority.

In the center of the highway an ambulance sat silent, its lights flashing red and yellow, fluttering off the delicate blanket of shattered glass.

"Officer what happened?" the lead woman asked.

"I'm not sure ma'am. I just arrived myself. Can you please move back?" He tried to herd the group but they wouldn't budge. Just then the doors of the ambulance burst open. A woman leapt from the back and ran franticly toward the crowd of stunned onlookers.

"I killed my babies!" she screamed. Her expensive yoga outfit was covered in blood. "Oh my God, I killed my babies!"

"Ma'am…." The officer started.

A stunned and haggard paramedic sprang from the ambulance and ran for the mother.

"No you don't understand my babies are dead!" The mother clawed at the woman executive in the stripped suit. "I did it. I did it. Oh my God I did it! I could feel the urge. When I was driving, I could feel it …"

"It's ok," the woman tried to comfort her, waiting for the paramedic.

"I knew it was going to happen but I couldn't stop it," the mother melted into tears. "It was like an urge in my head. Oh my God my babies …" The paramedic wrapped his arms around the shattered woman and led her back to the ambulance.

"Please people," the officer pleaded. "We need to clear the area. We have to clear the area."

More onlookers arrived and a news crew ran up from the tangle of abandoned cars. Egerton moved out of their way and tried to get back to his car.

"You're Dr. Simon Egerton?" a voice called. "You're Dr. Simon Egerton, the renowned roboticist and artificial intelligence expert," a woman broadcaster pointed and motioned to her crew.

"What?" Egerton wasn't sure what was happening.

"Start now," the broadcaster barked to her crew. "You're Dr. Simon Egerton," she said again. "You're here to testify for Dr. Happy in the Bok murder case."

"I don't know what you're talking about," Egerton replied and jogged awkwardly to his car.

"This is Bernadette Samuels for FNN at a truly horrific scene. A tragedy made even more mysterious by the presence of Dr. Simon Egerton, renowned roboticist and artificial intelligence expert who will be testifying for Dr. Happy in the Bok murder case."

* * *

"You've not been here more than an hour and you're already famous," Dr. Sellings Freeman opened the door and smiled.

"What's going on?" Egerton asked walking up to the house. "I was just …"

"I know. I know," Sellings held up his thin hand. "I saw it all on FNN. That Bernadette Samuels is a real piece of genetic engineering. Did you know her mother was also in the business? I think …"

"Sellings can I please come in? What's going …"

"Slowly Simon. Slowly." Sellings came out of the house into the barren front yard of the towering house. Heat waves from the ornamental rocks wafted up around them as if they were in an underwater inferno. "What do you think of my new home? Remarkable yes?" They were surrounded by an elegant over-architected suburban sprawl. Giant glamorous houses were perched along lazy curing avenues. The subdivision was on the edge of Maralinga Gardens and gave not a single clue that anyone lived in any of the houses. At the cross street of Anangu Way a senior-care bot walked a French bulldog in the hot sun.

"Who is Dr. Happy?" Simon asked. "What is the Bok murder case?"

"You see that bot right there?" Sellings pointed. "That is my only neighbor."

"What?"

"Yes. Yes. I know. It is strange, but I have seen no other neighbors. They built all of these houses for some housing boom that never happened. Can you call it a ghost town if no souls ever lived here? I should ask my neighbor," Selling waved to the bot as it disappeared into the empty sprawl. "Come Simon, let us go inside. I can't stand this heat."

"Who is Dr. Happy? What was that woman talking about?" Egerton asked as they stepped inside. The air was cold and dry.

"I'm Dr. Happy," Sellings replied with a smile.

"You?"

"Yes. It's all to do with the work I'm finishing up at Claxton Neuroscience. I haven't published anything yet. The business people don't think it's the right *time*. I do hate them." Sellings led Egerton through the dark and unfurnished house.

"But why Dr. Happy?"

Sellings sighed. "In some of the tests we stimulated the brain to produce isolated emotions in our subjects. It was quite fascinating. We were able to isolate guilt, shame, regret, loss, impulsivity … all of them. But Bernadette and FNN just heard I was stimulating emotions so hence … Dr. Happy."

"But why do they care about you and your work?" Egerton asked as they moved through a living room filled only with pristine wall-to-wall carpet. The harsh sun filtered through the drawn shades, giving the room a soft peach glow.

"I guess it could have been worse. I could have been Dr. Shame. That would have been bad." Sellings led Egerton through the kitchen and down to the basement.

"Sellings, really you have to tell me what's going on."

"FNN cares about me young Simon because of Edward W. Bok. I was getting to that … don't be so impatient. I was getting to that."

"Who is Edward Bok?" Egerton asked as they stopped in a windowless basement. It was cool and crammed with a mess of computers and broadcasting equipment.

"Bok was one of my old lab assistants at Claxton Neuroscience. He started working with us early in the experiment, but we had to let him go. He failed his third psych test," Sellings said as he woke up a bank of computers and a projector. "We couldn't have crazy people messing with people's brains."

"So Bok murdered someone?" Egerton asked. "That's what that news woman said."

"Yes," Sellings sighed as a 3D projection snapped to life. The back of the basement filled with a meticulous life-sized replica of a living room; on the floor was a dead woman. The details of the scene were perfect and unsettlingly. "To be more precise, Bok murdered this woman, Ruth Ashmore." Sellings pointed at the body. His voice sounded weary for a moment, almost sad. "And before that, the crazy nut shot her poor son, Tony, as he sat in his car out in front of her house." Selling switched the projection to show a replica of the front of the house, the car and the dead boy.

"What does this have to do with you?" Egerton asked. He walked toward the projection of the car, repulsed and curious.

"Bok didn't know the woman or her son," Sellings explained. "The investigator can't find any connection between them save for the fact that Bok murdered them both. He lived across the street here in Maralinga Gardens. He had never even met them. But there is no doubt that he killed them."

"Did Bok say why he did it?" Egerton was standing close to the dead boy, so close that he was worried he might catch the smell of the dead body.

"That is where I come in," Sellings flipped the projection back to the living room and the dead woman on the floor. This put Egerton with his legs in middle of the couch. "Bok says that yes he did indeed kill Ruth and Tony but that he can't be held accountable because he had no control of himself."

"He's pleading insanity?"

"Not exactly," Sellings answered. "It's not that simple. No, he's saying he had no control over his actions. That he knew he was going to do it, that he had felt the urge coming on all morning but at the moment before the murder he had no control, that he couldn't control himself, that in fact he had no free will to stop himself."

"That's ridiculous," Egerton replied, leaning over Ruth's body.

"Is it?" Sellings joined Egerton by the dead woman. "Because he's right. There is no such thing as free will. That's what I proved at Claxton. There is no neurological connection between the thoughts you think and the actions you take."

"But what about ..." Egerton wasn't buying it.

"It's true Simon," Sellings replied, his face growing tight and serious. "I have all the data. Free will is humanity's great delusion."

"Ok," Egerton held up his hand. "Let's say I believe you. What are people supposed to do now that you've told us all that we're deluded?"

All the color drained from Sellings face. He stepped close to Egerton, walking through the dead woman's body. "That's just it," he whispered. "I know what people will do. That's why I needed you to come here. I need your help. You have to help me stop it."

"Stop what?" Egerton was lost once again.

"I know what happens ..." Sellings started then stopped. "I know. They kill each other ..." he whispered. "... and it's all my fault."

* * *

"That's ridiculous," Egerton stepped away from his mentor, uncomfortable and worried he shouldn't have come.

"I wish it was," Sellings turned off the projection and moved back to Egerton. "But you saw that woman on the highway … the one who ran her car into that ridiculous sign. She killed her children. I tell you Simon, it's like a disease … the more people think about it … the more they try to control it … the brain begins to fall apart. People need the delusion to survive."

"Stop it."

"I drove her to do it," Sellings kept going. "It's true. While you drove here they looked into her past. Her name is Valerie Schwartz and until today she was the perfect wife and mother. No history of violence or mental illness … and yet today she killed her two children."

"That's crazy," Egerton shook his head. "You can't think that her accident has anything …"

"But she's not the first Simon … she's not the first. There have been twelve violent deaths in Maralinga in the past two days. It's a plague and it's only going to get worse." Sellings covered his face; his thin dry fingers shook with fear and shame.

"Sellings," Egerton tried to comfort his mentor. "This is crazy. There has to be another explanation …"

"There is no other explanation!" Sellings hands shot out and grabbed Egerton shoulders. "It was my pride. My father always told me my pride would ruin me." The old man's grip was surprisingly strong. "Once the FNN people found out Bok had worked for me, they called and asked my opinion. What could I tell them? I told them the truth but they didn't believe me. So they invited me to come here," he motioned to the empty basement. "To prove my theory. They said they would put me on TV, let me tell everyone what I'd found. Those idiots at Claxton couldn't stop me. That Samuels woman interviewed me here two nights ago."

"So that's what all this stuff is, for a TV interview?" Egerton still found it a little hard to look at Sellings.

"Yes," Sellings spat in disgust. "It was the first of two interviews. She just asked me stupid background questions that set up my work at Claxton. But it was enough. Most of Maralinga was watching and that's how it all started. Tonight is the final segment. Tonight they want me to talk about what I found … but I won't … I can't."

"So cancel the interview," Egerton replied. "Get out of here."

Sellings laughed pitifully. "Ha! If only it was that easy. I have to lie at my deposition too. It's in forty-five minutes. I have to lie. I'm ruined." Sellings hid his face again.

"How can I help?" Egerton was desperate to snap the old man out of his mania.

"I want you to tell FNN about your failure." Sellings answered quickly.

"My failure?"

"Yes. Everyone knows the trouble that you've been having with your work recently." Sellings eyes were frantic. "I couldn't ask you directly over the phone, my conversations are being recorded. Your robots … everyone knows you've got them thinking but you can't get them to be self-aware."

"Oh that," Egerton replied. "That's not a failure. That's just their development. We're making great progress …"

"But it *is* a great failure," Sellings shook as he spoke. "It's the same thing don't you see? It's free will. That's how Bernadette will see it. That's how everyone will see it. It's how they must see it. I'll testify that Claxton proves Bok to be a liar and he'll go to jail. You'll tell Bernadette that you cannot give your robots free will, that it is uniquely human, something only we humans have…."

"I see," Egerton replied after a moment.

"If you don't Simon more people will die."

"Alright, I understand," Egerton was exhausted and defeated. "But I won't lie for you Sellings. I'll tell them what I know but I won't lie."

"You don't have to," Sellings beamed with hope. "Bernadette and the crew will be over tonight for the broadcast."

"They just want to interview me?" Egerton asked.

"Yes just an interview," Sellings' mood lightened even more. "I need to go down to the court now but you should stay here. Thank you, Simon. Thank you so very much."

Ruth Ashmore lay dead on the floor of her bland living room. Egerton stood over her body and worried about the interview. Bok had shot the woman twice in the chest and then once in the forehead. Blood soaked the carpet and Egerton could see gore splattered on the wall.

"Tell me Ruth Ashmore did you know you didn't have free will?" Egerton asked the dead woman. "Nah you were just fine." Egerton switched the projection back to the front of the house and walked over to Tony. "Tony, did your mother ever tell you you didn't have free will?" Egerton chuckled at himself and thought of Sellings. He felt ridiculous and trapped. He looked at the poor dead boy and said, "I bet your mother never told you …" then stopped.

Inspiration fluttered across his mind with the speed of a camera flash and the force a hurricane. It was so simple. He understood how to give Jimmy his own free will. It was wonderful and simple and he felt ridiculous for not thinking of it before.

Egerton scanned the basement. He had to get out of Maralinga Gardens as quickly as possible. There was no way he could talk to FNN now. Frantically Egerton searched for his phone and dialed.

"Maralinga Rentals, this is Frances. If you have to be here, then you HAVE to be HERE. How can I help you?"

"Hello Francis, this is Simon Egerton."

"You didn't wreck the car did you?"

"No," Egerton smiled. "The car is fine. But Frances has the shuttle left yet? Today's shuttle?"

"What?" the girl was puzzled for a moment. "Oh no, it's still here." She looked away from the phone then came back. "It's just getting ready to go. Why?"

"Can you book me on it?" Egerton asked. "Something came up. I have to go back home right away. I have to be on that shuttle."

* * *

The beast of a car roared to life and clawed at the road when Egerton let it free. He raced down the empty suburban street, barely keeping control. Twice the car ran up onto the side walk and Egerton had to wrestle it back into the street. Just when he felt like he was getting the hang of the modified machine he saw the FNN news truck coming at him.

"No. No. No!" Egerton yelled as they passed him. Bernadette Samuels looked down at him, confused and then shocked.

Egerton panicked. He gunned the engine into a sharp turn. The car slid across the corner, knocking down the Anangu Way street sign. The news truck made a sharp u-turn and sped after him. Egerton couldn't let Bernadette interview him or get him on camera. He had to get out of Maralinga Gardens.

Launching out of the subdivision, Egerton aimed the car toward the shuttle port. FNN tried to close in but the burly engine left them in its dust. Just like earlier, the highway was mostly empty. The few cars Egerton did pass swerved to get out of his way when he tore past them. Nervously, he checked the rear cameras, waiting to see a police cruiser trying to over take him. But it was the front windshield that filled with the flashing lights of police cruisers and first responder vehicles. They blocked the highway completely. Fighting the car, Egerton got it to slow down in time to catch sight of a small gap in the vehicles and an officer waving him through, slowing him down.

Once again the surface of the highway was covered with a delicate and dangerous blanket of glass. Egerton's pulse quickened when he saw the tangle of smashed cars and the limp bodies being pulled from them. Two person EMT teams fought frantically to keep the living alive. A stoic officer covered a dead woman's face with his blast-resistant vest.

"You have to listen to me!" an executive pleaded with an EMT. "You don't understand …" his white shirt was sprayed with blood. One of his shirt cuffs was missing.

A nimble panic shivered through Egerton. Worried he might be contaminated by the accident's insanity he abruptly exited the highway.

Tidy office parks and executive meeting centers blurred past. Maralinga's business district was deserted. Egerton wondered if the few remaining executives were hiding out in their offices. The late day sun reflected off buildings' mirrored windows as if the power brokers were trying to send him coded messages.

Nearing the shuttle port, Egerton slowed the car not wanting to draw too much attention to himself. He had made it this far without getting pulled over and he didn't want to risk it.

He retuned the car to its petite parking lot outside the main concourse. Climbing from the car he saw the FNN news vehicle tearing towards him, reckless and at top speed.

Egerton sprinted into the concourse.

"Frances!" he yelled and ran over to the plump girl. "Were you able to do it? Did you get me on?"

"You wrecked the car didn't you?" she asked matter-of-factly. "Really I don't care. It's just, *you know*, my boss is going too …"

"I did scratch it against a street sign, but you can bill me for it," Egerton tried to calm down. He didn't want to scare the girl. "Were you able to get me on the shuttle?"

"You know that's *like* illegal right?" She fussed with her glasses. "But *you know* because my dad's on the council and all …"

Egerton leaned over the counter and kissed Frances on the forehead. "Thank you Frances. Thank you. Thank you. Thank you." Egerton slid her the rental keys and ran for the terminal.

"Sure but I …"

Bernadette Samuels and her news crew stormed through the door.

"This is Bernadette Samuels for FNN," she was broadcasting as they ran. "We are chasing renowned roboticist and artificial intelligence expert Dr. Simon Egerton who has refused us an interview about the Bok murder case."

"I have a reservation on the shuttle," Egerton said to the massive aboriginal security guard, trying not to sound out of breath. He handed him his passport.

"Listen old fella, you in big hurry inna? You reckon you are leaving?" the guard asked studying the papers.

"Something has come up and I need to get back to my work." Egerton looked over his shoulder and saw the light crowds clearing to make room for the FNN crew. "Am I in time?"

"Yiwi," the guard handed the passport back. "They been holding that big mudugor you, inna. We only got'em but one mudagar today … you gotta hurry up!" The guard ushered Egerton into the security area.

"It's unclear why the doctor is refusing us an interview," Bernadette panted into her mic. "Is there something he doesn't want us to know? Does he have some evidence in the Bok murder case that he doesn't want to get out? Dr. Happy is giving his first deposition right at this very moment."

The crew ran up against the guard.

"We need to get through," Bernadette pointed at Egerton as he moved out of the security area to the waiting shuttle.

"You don't have a reservation for the shuttle," the guard replied coolly, staring down at the sweating crew.

"No," Bernadette replied, leaning past the guard trying to keep an eye on Egerton's progress. "But we must get through. That man has crucial evidence in the Bok murder case."

"You white fella all the same … always in a big hurry," the guard smiled and shrugged. "But if you don't gott'em but reservation, been staying here, inna."

* * *

"Jimmy will you fix me and Dr. Freeman a drink?" Egerton asked.

"Sure thing," the bot replied cheerfully. "As I remember Dr. Freeman likes Traditional Pimms number one …"

"That would be lovely Jimmy," Sellings replied with an astonished smile. "Thank you for remembering."

"No problem at all," Jimmy answered then set about making the drinks.

Egerton and Sellings sat in Egerton's cramp apartment in the clog of stations that ringed Earth.

"Simon, he really is a silly robot," Sellings said watching Jimmy's small child-like body work at the bar.

"I know," Egerton grinned, pleased Jimmy. "He's a funny little guy but he keeps me company …" he trailed off then added, "… he's the first one, you know. Jimmy will do down in history."

"Yes, yes, yes … I read your paper and saw your telecast," Sellings replied, unable to hide his annoyance. "*Teaching Free Will as an Alternative to Metathought Safeguards* … you couldn't have thought of a catchier title?"

"I hope you're not too mad at me." Egerton was sheepish. "I had to leave Maralinga. I didn't want to leave you like that but I couldn't lie and when I finally figured it out I just couldn't …"

"I understand," Selling didn't hide his dissatisfaction. "I was more confused when I heard. Bridgette was livid."

"I saw Bok got life in prison," Egerton said, trying to move on.

"Yes," Sellings sighed. "My lie locked up a murderer, ended my career and saved Maralinga Gardens."

"Your career isn't over …"

"Stop," Sellings waved his hand in front of his face. "Let an old man be bitter if I want to. Yes. Yes. I still have work but it's not the same."

"But seriously Sellings," Egerton had rehearsed the next words for several weeks. "I do have you to thank for helping me. If I hadn't come to Maralinga …"

"Shut up Simon!" Sellings interrupted. "I don't care about your speech. Just tell me what you did." He pointed at Jimmy. "What did you do?"

"I just told him he was free," Egerton laughed. "That's it. It was simple. I taught him he had free will. That's it really. Then everything changed."

Sellings sucked in air with astonishment. "That's it? How funny…."

There was a gentle pause in the conversation as both men played around with their own thoughts.

"There's just one thing," Egerton interrupted as Jimmy returned with the drinks.

Sellings took a sip of his Pimms and replied, "What's that?"

"Well …," Egerton paused. "I think Jimmy is developing a soul."

"Really?" Sellings stared at the cheerful little bot and the cheerful little bot stared back.

"How's your drink Dr. Freeman?" Jimmy asked.

¹ Department of Wildlife

² Rice cooked in coconut milk, served with peanuts and anchovies

³ Traditional healers

⁴ spirit

⁵ guardian spirits that resides in trees, ant hills, caves, riversides, and in stone formations

⁶ skill and knowledge

⁷ mouse deer

⁸ brother, term of endearment

⁹ blowpipe with poisoned dart from the Ipoh tree

¹⁰ Malay machete

¹¹ (lit.) Ali's stick. *Eurycoma longifolia* is a flowering plant with reputation as an aphrodisiac

¹² a tropical rainforest tree

¹³ sweet potato

¹⁴ *Symphalangus syndactylus*, a tailless arboreal gibbon

¹⁵ natives

¹⁶ magical power

¹⁷ traditional customs and law

[18] (lit.) rat island

[19] Straits-born Chinese women who are partially assimilated to local culture speaks a blend of Chinese, Malay, and English

[20] *Neobalanocarpus heimii*, type of hardwood

[21] honorific non-hereditary title conferred by head of state

[22] (slang) fix

[23] (slang) thoroughly

[24] (slang) fix

[25] (slang) thoroughly

[26] Straits Chinese delicacies (edible)

[27] Corresponding author

Notes

Aldiss Brian and David Wingrove. 2001. *Trillion Year Spree*. London: House of Stratus.

Ashley, Michael. 2000. *Time Machines: The Story of the Science-Fiction Pulp Magazines from the Beginning to 1950*. Liverpool: Liverpool University Press.

Asimov, Asimov and Stanley Asimov ed. 1995. *Yours Isaac Asimov: A Lifetime of Letters*. New York: Doubleday.

Asimov, Isaac. 1942. "Runaround." *Astounding Science Fiction*. March.

Asimov, Isaac. 1967. *Is Anyone There?* New York: Doubleday Books.

Asimov, Isaac. 1977. *"Our Intelligent Tools."* American Airlines.

Asimov, Isaac. 1981. *Asimov on Science Fiction*. New York: Avon.

Baker, Kelley. 2009. *The Angry Filmmakers Survival Guide*. Portland: Booksurge.

Baxter, John. 1970. *Science Fiction in Cinema*. New York: Paperback Library.

Benford, Gregory and Elizabeth Malartre. 2007. *Beyond Human-Living with Robots and Cyborgs*. New York: Forge.

Bleecker, Julian. 2009. *Design Fiction: A Short Essay on Design, Science, Fact and Fiction*. Near Future Laboratory. http://drbfw5wfjlxon.cloudfront.net/writing/DesignFiction_WebEdition.pdf.

Bostrom, Nick. 2005. "A History of Transhumanist Thought." http://www.nickbostrom.com/papers/history.pdf.2006.

Brady, John. 1981. *The Craft of the Screenwriter*. New York: Simon and Schuster.

Brooks, Michael. 2009. *13 Things That Don't Make Sense*. New York: Vintage.

Clark, Arthur. 1999. *Greetings, Carbon-Based Bipeds!* New York: St Martin's Press.

Close, Frank. 2009. *The Void*. New York: Sterling.

Creative Science Foundation. 2010. "1st Workshop on Creative Science (Kuala-Lumpur, July 2010)." *Creative Science Foundation*. http://www.creative-science.org/.

David, Peter. 2006. *Writing for Comics with Peter David*. Cincinnati: Impact.

Doctorow, Cory and Karl Schroeder. 2000. *The Complete Idiot's Guide to Publishing Science Fiction*. Indianapolis: Alpha books.

Doctorow, Cory. 2004. "Rise of the Machines." *Wired*. July.

Doctorow, Cory. 2005. "I, Robot." *InfiniteMatrix.net*.

Doctorow, Cory. 2006. "I, Row-Boat." *Flurb#1*.

Doctorow, Cory. 2007. "Free Data Sharing Is Here to Stay." *The Guardian*. September 18.

Doctorow, Cory. 2007. "Introduction to I, Robot." *Overclocked*. Philadelphia: Running Press.

Doctorow, Cory. 2008. *Little Brother*. New York: TOR.

Doctorow, Cory. 2010. "About Cory Doctorow." *Craphound*. http://craphound.com/?page_id=1638.

Dourish, Paul and Genevieve Bell. 2009. *Resistance is Futile: Reading Science Fiction and Ubiquitous Computing*. http://www.dourish.com/publications/2009/scifi-puc-draft.pdf.

Egerton, Simon, Victor Callaghan, and Graham Clarke. 2008. "Using Multiple Personas in Service Robots to Improve Exploration Strategies when Mapping new Environments." *Intelligent Environments, 2008 IET 4th International Conference*.

Egerton, Simon, Victor Callaghan, Victor Zamudio, and Graham Clarke. 2009. "Instability and Irrationality: Destructive and Constructive Services within Intelligent Environments." *Intelligent Environments 2009—Proceedings of the 5th International Conference on Intelligent Environments—Barcelona 2009*.

Eisner, Will. 1985. "Preface." *Comics and Sequential Art*. Tamarac: Poorhouse.

Eisner, Will. 1996. *Graphic Storytelling and Visual Narrative*. Tamarac: Poorhouse.

Eisner, Will. 2001. *Will Eisner's Shop Talk*. Milwaukee: Dark Horse Comics.

Encyclopedia Britannica. 2008. *The Britannica Guide to the 100 Most Influential Scientists*. London: Running Press.

Evans, Russell. 2010. *Stand-Out Shorts*. Burlington: Focal Press.

Field, Syd. 1979. *Screenplay*. New York: Dell.

Fox, Gardner. 1961. "Inside the Aton." *Showcase #34*. New York: DC Comics.

Goulart, Ron. 2001. *Great American Comic Books*. Lincolnwood: Publications International, Ltd. Gresh, Lois and Robert Weinberg.2002. *The Science of Superheroes*. Hoboken: Wiley.

Gresh, Lois H. and Robert Weinberg. 2005. *The Science of Supervillians*. Hoboken: Wiley.

Internet Movie Database. 2010. "All Time Box Office: USA." http://www.imdb.com/boxoffice/alltimegross.

Internet Movie Database. 2010. "Moon." http://www.imdb.com/title/tt1182345/.

Internet Movie Database. 2010. "Sibyl." http://www.imdb.com/title/tt0075296/.

Kriz, Sarah, T.D Ferro, and J.R. Porter. 2010. "Fictional Robots as a Data Source in HRI Research: Exploring the Link between Science Fiction and Interactional Expectations." *In Proceedings of the 19th IEEE International Symposium in Robot and Human Interactive Communication*. Ro-Man.

Loke, Kar-Seng and Simon Egerton, 2010. "Automated Eye on Nature (AEON) and The Were-Tigers of Belum." *Workshop Proceedings of the 6th International Conference on Intelligent Environments*: pp. 261–70.

McCloud, Scott. 1993. *Understanding Comics*. New York: Kitchen Sink Press.

Moore, Alan. 2008. *Alan Moore's Writing for Comics*. Rantoul: Avatar Press.

Perkowitz, Sidney. 2004. *Digital People*. Washington, D.C.: Joseph Henry Press.

Perkowitz, Sidney. 2007. *Hollywood Science*. New York: Columbia Press.

Perkowitz, Sidney. 2010. "Sidney Perkowitz Bio." http://www.sidneyperkowitz.net/about/bio.php.

Schneider, Steven Jay. ed. 2009. "Introduction." *Sci-Fi Films You Must See Before You Die*. Hauppauge: Barrons.

Shedroff, Nathan and Chris Noessell. 2012. *Make It So: Learning From Science Fiction*. Brooklyn: Rosenfeld Media.

Shelley, Mary. 1831. Introduction. *Frankenstein*.

Steele, Alan. 1992. *Hard Again*. New York Review of Science Fiction.

Wells, H.G. 1933. Preface. *The Scientific Romances*.

Zizzi, P.A. 2008. "I, Quantum Robot: Quantum Mind Control on a Quantum Computer." *Journal of Physics Conf. Ser.*, 67.

Author Biography

The future is Brian David Johnson's business. As a futurist at Intel Corporation, his charter is to develop an actionable vision for computing in 2021. His work is called "future casting"—using ethnographic field studies, technology research, trend data, and even science fiction to create a pragmatic vision of consumers and computing. Johnson has been pioneering development in artificial intelligence, robotics, and reinventing TV. He speaks and writes extensively about future technologies in articles and scientific papers as well as science fiction short stories and novels (*Fake Plastic Love and Screen Future: The Future of Entertainment, Computing and the Devices We Love*). He has directed two feature films and is an illustrator and commissioned painter.